Cambridge Monographs in African Archaeology
52
Series editor: John Alexander

The Middle and Later Stone Ages in the Mukogodo Hills of Central Kenya

A comparative analysis of lithic artefacts from Shurmai (GnJm1) and Kakwa Lelash (GnJm2) rockshelters

G-Young Gang

BAR International Series 964
2001

Published in 2016 by
BAR Publishing, Oxford

BAR International Series 964

Cambridge Monographs in African Archaeology 52

The Middle and Later Stone Ages in the Mukogodo Hills of Central Kenya

ISBN 978 1 84171 251 2

BAR Publishing is the trading name of British Archaeological Reports (Oxford) Ltd.
British Archaeological Reports was first incorporated in 1974 to publish the BAR
Series, International and British. In 1992 Hadrian Books Ltd became part of the BAR
group. This volume was originally published by Archaeopress in conjunction with
British Archaeological Reports (Oxford) Ltd / Hadrian Books Ltd, the Series principal
publisher, in 2001. This present volume is published by BAR Publishing, 2016.

Printed in England

BAR
PUBLISHING

BAR titles are available from:

BAR Publishing
122 Banbury Rd, Oxford, OX2 7BP, UK
EMAIL info@barpublishing.com
PHONE +44 (0)1865 310431
FAX +44 (0)1865 316916
www.barpublishing.com

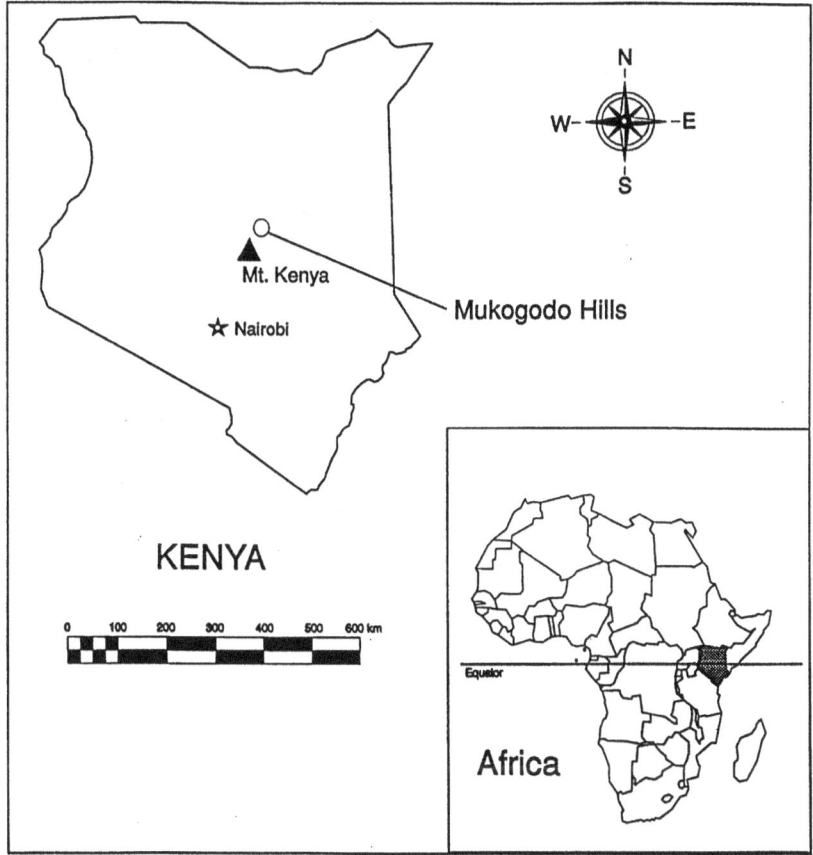

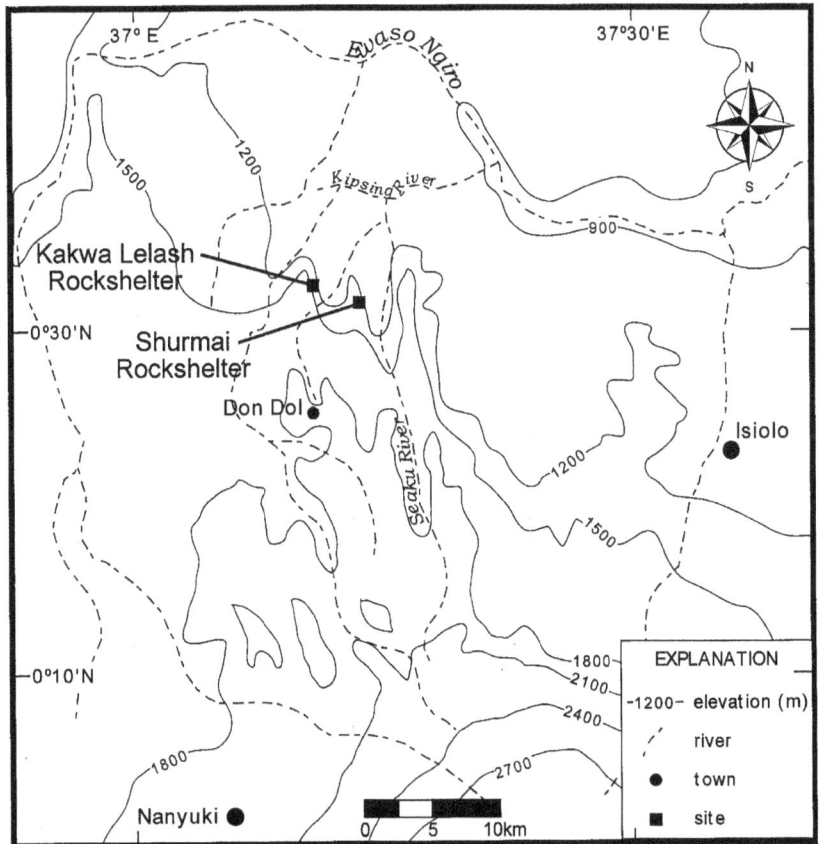

Table of Contents

TABLE OF CONTENTS (continued)

TABLE OF CONTENTS (continued)

TABLE OF CONTENTS (continued)

LIST OF FIGURES

LIST OF TABLES

Preface and Acknowledgements

This monograph compares lithic materials excavated from Shurmai (GnJm1) and Kakwa Lelash (GnJm2), two deeply stratified dry rockshelters located in the Mukogodo Hills of the Laikipia and Isiolo districts of Central-North Kenya. The analysis forms a key part of the reconstruction of the sequence of human occupation in the region back to the Middle Stone Age. Lithic components from these two sites are analyzed and compared in terms of their raw material, techno-morphological attributes, function, and styles. The analysis reveals that there are both difference and similarities in raw material procurement strategies, morphological attributes of artifacts, and stylistic variations between the Middle and the Later Stone Age occupations at these two archaeological sites. However, there is little difference in technology and the functional attributes of the tools through time. The two sites indicate that from the Middle to the Later Stone ages the changes in the lithic industry were gradual.

This monograph emerged out of my dissertation done in the Department of Anthropology of Texas A&M University. Its writing would have been impossible without the help, cooperation, and encouragement of many people. I want to thank the administrative staff of the Department of Anthropology including Joyce Bell, Barbara Spear, Karen Taylor, and Becky Jobling. They gave me many hugs, laughter, and enormous support. Concerning flint knapping, my first and most important acknowledgment is to Dr. Errette Callahan, Director of Piltdown Production in Lynchburg, Virginia, and one of the best knappers in America. He made understand and correctly read lithics and helped me solve many technical problems related to this monograph. I also would like to thank William Dickens, a TAMU graduate student and one of the finest flint knapper in Texas. Thanks also to Drs. John E. Dockell, Harry J. Shafer, Ralph Solecki, and Rose Solecki all of the Department of Anthropology. My knowledge of lithics is in large measure the product of their careful teaching.

Much of the analysis upon which this monograph is based was done in laboratory of the Division of Archaeology of the National Museums of Kenya in Nairobi, Kenya. I want to thank the warm, friendly and knowledgeable staff of the Division and especially Dr. Karega-Munene, Division Head at the NMK. During my stay in Nairobi they always spared time from their busy schedules to help me with my research.

I would like to thank my American family, Dawn LaBelle Marshall and Andrew Marshall for all the weekends they invited me to their house and fed me. They always made me relax from my work and gave me enormous help. They taught me the traditional Midwestern American way of life. I also want to thank to my committee, Drs. Fred Smeins, Lee Cronk, and David Carlson. However, most of all, I want to give my special thanks to my professor Dr. D. Bruce Dickson who is committee chair, mentor, colleague, and my best friend. I finally would like to thank my parents, Young-Whan Gang and Yong-Soon Yi; my sisters, G-Hyon, G-Hee, Nam-Hee; my brother Young-Jae, and my little two nephews, Tae-Ho and Min-Ghi. Their endless love and support have carried me through these last six and half years.

Chapter I
Introduction

For the last five years a team from the Department of Anthropology at Texas A&M University and the National Museums of Kenya has conducted a long term research project aimed at reconstructing the ethnoarchaeological and geological and paleoenvironmental history and prehistory of the Mukogodo Hills region of the Laikipia and Isiolo districts of Kenya during the Late Quaternary period. This research was initiated with a four-week site reconnaissance during the dry season of 1992 and has continued to the present time with excavation work and further site survey.

Over a four-week period during the dry season of 1993, preliminary test excavations were conducted in the Shurmai Rockshelter (GnJm 1) (Figure 1). During the dry season of 1994 the initial test excavations were expanded and a two-by-two meter *sondage* was excavated to bedrock. These two seasons of excavation have provided a clear stratigraphic profile of the northern end of the Shurmai Rockshelter. A preliminary thermoluminescence date indicates that the earliest deposits at the site date to the Middle Stone Age.

When bedrock was reached in the Shurmai Rockshelter in 1994 without encountering any evidence of a hominid presence beneath the Middle Stone Age occupation layer, the excavation team shifted to another rockshelter, called Kakwa Lelash (GnJm 2), hoping to find evidence of earlier occupation (Figure 2). Over a six-week period in 1995 a deep two-by-two meter *sondage* in the deposits at the site was excavated. Both stratigraphic and soil analyses reveal that the earliest occupations at Kakwa Lelash site belong to the Later Stone Age. The two rockshelters produced abundant lithic materials. The present work, which is a detailed analysis of that lithic trial, was undertaken as part of the Mukogodo Hills Project.

The African prehistoric sequence consists of three periods: Earlier, Middle, and Later Stone ages. These periods or ages can be distinguished from each other on the basis of changes in stone tool technology. The Earlier Stone Age (hereafter ESA) begins with the first appearance of the stone artifacts ca. 2.9 million years ago and ends with the disappearance of large bifacial cutting tools called handaxes and cleavers ca. 200,000 years ago. From about 1.6 to 1.5 million years ago, the stone artifacts recovered from African archaeological sites begin to show an increase in retouched stone artifact forms. The following Middle Stone Age (hereafter MSA) is characterized by flakes, blade flakes, points, and prepared cores. Later, about 40,000 years ago, this flake industry is replaced by microlithic stone artifacts. The appearance of microlithic industry signals the beginning of the Later Stone Age (hereafter LSA). The date of the end of the Later Stone Age has not yet been established. However, in general, it is believed that the Later Stone Age ends about 3000 B.P. (Robertshaw 1995).

In general, archaeologists in Africa have focused on the Earlier Stone Age. Perhaps this is because the Earlier Stone Age has produced key fossils and other spectacular evidence of hominid evolution in the eastern and southern parts of the continent. Research on the Later Stone Age has also been extensively developed. Perhaps this is because the study of the Later Stone Age has been facilitated through the use of ethnographic analogies with modern peoples and has been enhanced by the development of the radiocarbon dating method. Compared to research done on the Earlier and Later Stone ages, study of the Middle Stone Age remains underdeveloped and beset with many difficulties relating to terminology and a paucity of extensive regional research.

Recently however, the study of the Middle Stone Age has moved to the center of African prehistoric studies as it has been recognized as the period when anatomically modern humans (*Homo sapiens sapiens*) first appear in prehistory. Archaeological data suggest that anatomically modern humans evolved in the later middle Pleistocene overlapping the Middle Stone Age there. The development of more refined methods of dating the period of the Middle Stone Age has spurred research as well. Even so, research on the Middle Stone Age in East Africa remains in its infancy when compared to that of the South Africa.

Shurmai (GnJm 1) and Kakwa Lelash (GnJm 2) Rockshelters

Recent excavations of the deeply stratified Shurmai (GnJm 1) and Kakwa Lelash (GnJm 2) rockshelters in Central-North Kenya have revealed a sequence of human occupation extending back from the present to the Middle Stone Age. The Shurmai Rockshelter in particular is unique because it contains deposits of both the Middle and Later Stone ages, while the Kakwa Lelash Rockshelter consists of Later Stone Age and recent deposits. The transition from the Middle to the Later Stone ages in East Africa has been seen by some scholars as an abrupt, and by others as a gradual, transition. As one of our rockshelters contains both MSA and LSA deposits, it can help our understanding of the transitional period. In addition, the LSA lithic industries recovered from both sites can be compared with each other and with MSA lithic industries.

Unfortunately, the study of rockshelters has both advantages and disadvantages (Jelinek 1976). In terms of advantages, rockshelters are characterized by their preservation of a stratigraphic succession of restricted cultural elements. But, the sites are totally lithic sites, which do not produce human remains as well as any other cultural elements such as faunal and floral remains. Although the two rockshelters produced faunal remains, the remains were too small to be compared and to give information about the past environment. In addition, unlike an open site, which contains hearth or other features *in situ*, lithic sites do not contain any functional features. As a result, it is very difficult to distinguish the function of the artifacts. Due to the nature of the two rockshelters, reconstructing the subsistence system of the period

Figure 1. Shurmai Rockshelter (GnJm 1).

Figure 2. Kakwa Lelash Rockshelter (GnJm 2).

becomes not possible. Thus, the present study largely focuses on the lithics recovered from the two sites rather than floral and faunal remains.

Research Goals

The purpose of the present study is to understand the changes in lithic technology from the MSA to LSA. In order to understand this goal, Stone Age lithic materials excavated from Shurmai (GnJm 1) and Kakwa Lelash (GnJm 2), two deeply-stratified dry rockshelters, archaeological sites located in the Mukogodo Hills of Laikipia and Isiolo districts in Kenya are analyzed and compared. During this analysis, it is found that the two sites produced different raw material types and different lithic morphology. As the two sites belong to different cultural sequences, MSA and LSA, it is hypothesized that the two sites practiced different raw material procurement strategies and lithic technology would be changed through time.

Methodology

There are four primary factors that influence or limit the shape that a stone artifact may assume (Whittaker 1994:270): (1) the raw material out of which the tool is formed, (2) the technology used in its manufacture, (3) the function of the object, that is, the purpose or purposes to which it is to be put, and (4) the style or characteristic mode of tool form current among the people making the object. All four of these factors will be systematically considered in comparing the lithic assemblages from the various components at the Shurmai and Kakwa Lelash rockshelters.

Raw Material Analysis

Preliminary analyses of the lithic collections from the sites indicate that the occupants of Shurmai and Kakwa Lelash rockshelters made use of four primary types of raw materials: the most common of which is fine-grained basalt, the next most frequent are chert and large grained milky quartz, respectively, and the least common obsidian. The type of raw material used in stone working at the two shelters will be undertaken on the basis of the distance of the source area from the sites.

Analysis of the Technology of Manufacture

In analyzing the lithic collections from the two sites particular attention is paid to discerning (1) the stage of manufacture exhibited by each object, and (2) the techniques of manufactures in use during the formation of the various occupations at the sites. Examination of the collections indicates that hard and soft hammer percussion techniques were practiced during all occupations. In the lowest stratum at Shurmai the MSA lithics were evidently produced in large measure using hard and soft hammer percussions and disc core techniques. Flakes characteristic

of pressure flaking have been recovered from the highest strata at both sites.

Style

From the perspective of cultural interpretation style is the most important of the four factors. Therefore, particular attention is paid to formal analysis and description of the individual tools and artifacts recovered from the two sites. It is my view that stylistic variations represent specific choices by artisans from a broad spectrum of equally viable alternative ways of achieving the same end, and that style should be interpreted on the basis of functional consideration. Therefore, in the present study the styles of formal tools are analyzed based on the isochrestic approach. The definition of this approach is provided in the Chapter IV.

Function

Despite decades of concerted scientific attention to the problem, discerning the function of stone tools from archaeological sites remains problematical. This is particularly true of tools manufactured and used by hominids that were not anatomically modern. After all, we can never be certain that the lack of physical correspondence between these early hominids and ourselves does not also reflect profound cognitive and behavioral differences. Therefore, the use of direct analogies with modern tools to interpret ancient objects from the ESA or the MSA may not be warranted. For this reason, functional interpretation of the lithics from earliest components of the Shurmai and Kakwa Lelash rockshelters is conservative and minimalist in spirit, and based exclusively on (1) measurements of edge angles following the method of Wilmsen (1969), and (2) macroscopic and microscopic analysis of edge wear in the manner of Odell (1980).

Research Procedure

The Shurmai Rockshelter (GnJm 1) produced 4,782 stone artifacts, while Kakwa Lelash (GnJm 2) yielded 7,862 stone artifacts. All lithic artifacts (12,644) recovered from the two rockshelters are categorized as flakes, tools, cores, or debris. The lithic assemblage recovered from the Shurmai Rockshelter consists of 1,467 unmodified flakes, 100 cores, 79 tools, 1,306 fragments, and 1,830 pieces of debris. Of these, unmodified flakes consist of 843 complete and 624 broken flakes. The lithic assemblage recovered from the Kakwa Lelash Rockshelter is composed of 1,067 unmodified flakes, 587 of which were complete and 480 of which were broken. In addition, there were 32 cores, 26 tools, 1,726 fragments, and 5,011 pieces of debris. Only unmodified flakes, cores, and tools are analyzed in the present study.

The present dissertation consists of seven chapters. In Chapter I, I introduce the study. In Chapter II, I discuss the Paleolithic period in Africa focusing on the characteristics of African Stone ages in general and the Middle Stone Age in particular. Dates of the MSA, related hominids of the period, the definition of the MSA, and the LSA are provided. Although the present study dropped the local terms for the MSA, in order to help understanding the previous research, some of these local terms are explained.

In Chapter III, the environmental background and ecological features of my study area are discussed. The previous research history and the excavation procedures undertaken for last three years are described as well. Particular attention is given to the geological and physiographical setting of the two sites and their influence on the lithic industries there.

In Chapter IV, the methods of analysis used in the present study are discussed and the specific terms used in the study are defined. Chapter V is an account of general mechanics as applied to stone to produce flakes as well as mechanism of mechanic fracture and raw materials preferred by the stone knapper. In the final section of the Chapter V, the techniques found to have been used at the two sites are discussed.

In Chapter VI, I analyze, compare, and contrast the lithic materials recovered from the two rockshelters in numerical and statistical terms. In this chapter I discuss how the lithic industries of the study area changed and how occupants of the two sites probably acquired the materials they needed. This chapter focuses on the raw material, techno-morphological, stylistic, and functional analyses. In the concluding chapter, Chapter VII, I provide the results of analyses related to raw material procurement strategies, techno-morphology, style, and function of lithic materials recovered from the two sites. Based upon my analysis, future research is suggested.

Chapter II
Previous Research of the Study Area, Shurmai (GnJm 1) and Kakwa Lelash (GnJm 2) Dry Rockshelters

Politically, East Africa consists of the modern nations of Kenya, Uganda, Tanzania, Ethiopia, Somalia and parts of Central Africa such as Zambia, Malawi, and northern Mozambique (Robertshaw 1995:57). East Africa is a land of contrasts due largely to variations in altitudes caused by tremendous upheavals of the earth's crust and intense volcanic activity. Most of the countries lie at between 2,000 and 3,000 meters above sea level (Phillipson 1977:1). Kenya in East Africa is bordered on the east by both the Republic of Somalia and the Indian Ocean. The Republic of Tanzania borders Kenya in the south, Uganda on the west. To the north and northeast, Kenya is bordered by the republics of Sudan and Ethiopia, respectively.

Physiographic Features and Geological Setting of Kenya

Kenya covers an area of 582,646 square km (2,240,960 square miles) and is roughly the size of France. This is a very large region, which contains a great variety of physical features and physical environment (Ojany 1971:137). Kenya can be subdivided into four broad physical geographic regions: the Coastal Plain, Arid Low Plateau, Kenya Highlands and Rift Valley, and Lake Victoria Borderlands (Soja 1968:Figure 1) (Figure 3).

The Kenya Highlands and the associated Rift Valley are components of the East African Rift Valley, which is the most spectacular landscape feature of East Africa. The Rift Valley stretches from the Jordan Valley to the Zambezi and includes the Red Sea and the Gulf of Aden. This valley averages 40 miles across and is an area of internal drainage. Its floor lies about 1,500 feet below the general level of the surrounding countries (Cole 1963:35-37). As the Rift Valley profoundly affects the ecological and physiographical makeup of Kenya, a detailed description of its formation is provided here.

The rigid outer layer of the earth, including the crust and upper mantle consists of numerous individual segments. These segments are called plates; the earth's rigid outer layer is called lithosphere. Below the lithosphere there is a hotter and weaker zone. This zone is known as the asthenosphere. Plates move as units over the material of the asthenosphere and interact in various ways producing earthquakes, volcanoes, mountains, and the crust itself. This movement of plates is called plate tectonics. The earth's eight major lithospheric plates are the Eurasian, Indo-Australian, Pacific, Antarctic, Nazca, South American, North American, and African plates. In general tectonic activities occur along the boundaries of these planes. Plate boundaries consist of three types: divergent, convergent, and transform boundaries. When plates move apart, resulting in upwelling of material from the mantle to create a new sea floor, it is called divergent boundaries.

When plates move together causing one of the slabs of lithosphere to be consumed into the mantle, it is called convergent boundaries. If plates slide past each other without creating or destroying lithosphere, they are called transform boundaries. When the plates move apart from the area of upwelling, the broken slabs of the lithosphere are displaced downward, then create down faulted valleys. These valleys are called rift or rift valleys. As the spreading continues, the rift valleys lengthen and deepen. Finally, they extend into the ocean. At this final stage, the valley becomes a narrow linear sea with an outlet to the ocean. The East African Rift Valleys were produced by the initial stage in the breakup of a continent. Large volcanic mountains such as Kilimanjaro and Mount Kenya reflect extensive volcanic activity (Tarbuck and Lutgens 1993:460-465; White and McKenzie 1989:62-71). The Kenya Highlands are products of the rifting history in the area.

The Kenya Highlands is a series of high plateaus and volcanic surfaces lying between the elevations of 1,300 and 3,400 meters. The heart of this region is a portion of the eastern Rift Valley, which contains a number of small lakes, including Nakuru and Naivasha, as well as several volcanic cones and plugs. The escarpments of the eastern Rift Valley hem in the central part of the Rift Valley and form the edges of the Aberdares Range on the east, Mau Hills and the Kericho highlands, and the Uasin Gishu Plateau on the west. Several volcanic and non-volcanic high plateaus such as Laikipia and Trans-Nzoia are fanned off from the central portion of the Highlands (Soja 1968:6).

Geologically the Highlands are very complex. The basements rock systems, which form the underlying geological foundation, include granites, gneisses, schists, migmatites, granulites and quartzites. All of these rocks are metamorphic in character and date back to the Pre-Cambrian period. Approximately 20 percent of the total area of the Highlands is composed of these metamorphic rocks. The remaining 80 percent are Tertiary to Recent volcanic rocks. These volcanic rocks, which are closely connected with tectonic activity, consist of two types. They are plateau lavas and central type volcanism. Plateau lavas originate from fissure eruptions. These fissure eruptions produced the majority of the recent volcanic rocks in the Highlands and have given rise to lava plateaus such as the Kinangop Plateau, the Laikipia Plateau, and the Athikapiti Plains. The central type volcanism produced the vast amount of volcanic material, which was deposited on top of the plateau lavas. As a result, central volcanism is closely related to the uplands such as the high volcanic piles of Mount Kenya, Kinangop, Sattima in the Mount Aberdares, Mount Longonot, Eburru, the Menengai Cauldron, and Mount Elgon. In terms of rock composition, areas related to central type volcanism were influenced by the nature of the associated eruption. Therefore, they produce various types of rocks such as nephelines, basalts, trachytes, agglomerates, and tuffs. It is believed that the origins of this central volcanism are associated with crust disturbances during the final stages of the formation of the Rift Valley (Odingo 1971a:1-4).

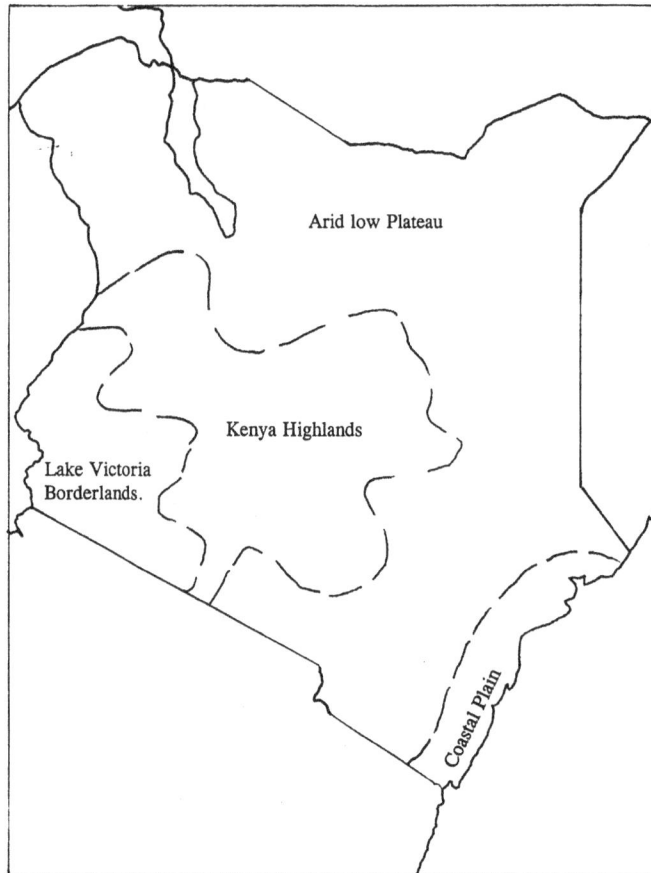

Figure 3. Physical Geographical Regions of Kenya.

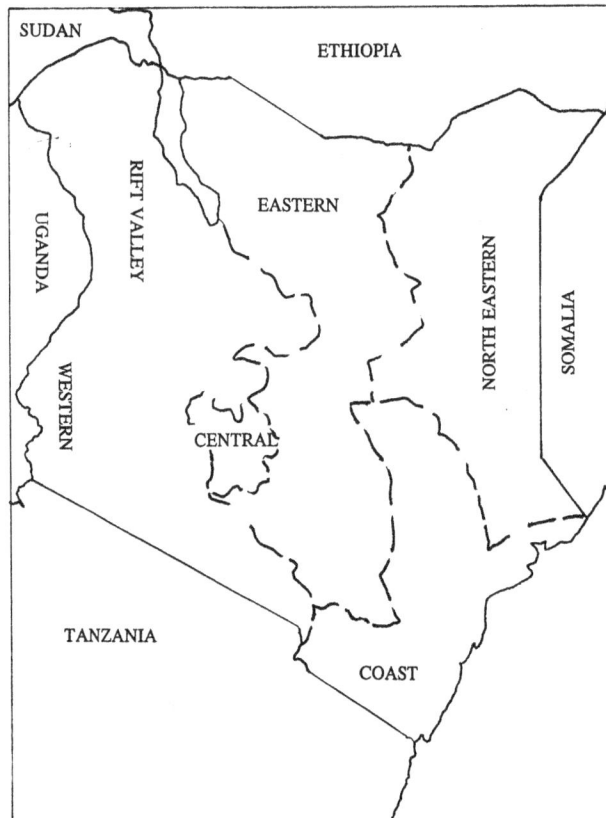

Figure 4. The Location of Kenya and Its Provinces.

According to Odingo (1971a:4-5), the Highlands may be divided into three subregions. They are the Highlands to the east of the Rift Valley, the Highlands to the west of the Rift Valley, and the Rift Valley itself. With the exception of the pre Cambrian basement rocks, the Highlands to the east of the Rift Valley are dominated by Tertiary to Recent volcanic lavas. The Laikipia Plateau is located within these eastern Highlands. The feature is a great lava Plateau area located between the elevation of 1,600 and 2,200 meters above the sea level (Flury 1988:265).

This great lava plateau is hemmed in to the west by the Aberdare Mountain range, to the southeast by Mount Kenya, and to the northeast by the basement systems of the Mukogodo Hills, which is metamorphic in character. The Laikipia Plateau consists of flat, gently rolling topography cut through by the various rivers, including the Narok and the Ewaso Ngiro and their tributaries (Gregory 1896:146; Odingo 1971a:8). Gregory (1896:146), an early visitor, found the plateau area reminded him of Wales.

The Location of Mukogodo Hills Regions of North-Central Kenya

Administratively, Kenya is divided into seven provinces: Western, Nyanza, Rift Valley, Central, Eastern, North Eastern, and Coast (Figure 4). Of these provinces, the drier and hotter parts of the northern and southern Rift Valley Province includes the districts of Turkana, West Polot, Baringo, Samburu, parts of Laikipia, as well as parts of Nakuru, Narok, and Kajiado districts (Ominde 1971:150, Figure 13.2).

Laikipia District belongs to the Rift Valley Province and bordered on the southeast by Central Province and on the northeast by Eastern Province. The district consists of four divisions: Ngarua in the northwest, Rumuruti in the southwest, Mukogodo in the northeast, and the Central divisions in the southwest and center (Flury 1988:Figure 1).

The Mukogodo Hills are located in the Mukogodo Division on the eastern part of the Laikipia Plateau. The Mukogodo Hills are characterized by rugged range of low mountains situated between the forested slopes of Mount Kenya the savanna grassland and deserts of northern and eastern Kenya. The elevations of these hills vary ranging from 1,000 to 2,100 meters above sea level. The eastern and western segments of the hills consist of a series of prominent peaks. The Seiguu River floors a deep valley between these two rows of hills. The eastern row of hills drops steeply to a flat, sandy plain, while the western row of hills is paralleled by a series of ridges that continue west into Mumonyot territory. Two rows of hills meet in the south where there is a large, open, grassy plain called Anandanguru (Cronk 1989:47). Anandanguru plain is drained by the intermittent Tol and Norgonono rivers that merge into the Kipsing River. These three rivers are all part of the Ewaso Ngiro catchment of northeastern Kenya (Dickson and Kuehn 1997:2).

A series of inselbergs, or erosional rock remnants, is the dominant topographic feature in the dry valley between the mountain. The Mukogodo Hills are made of granitic gneisses (Dickson, Kuehn, and Cronk 1995:1). The summit levels above 2,100 meters are suggested as remnants of either the Cretaceous or Sub-Miocene erosion surfaces (Shackleton 1946:44). In contrast to the rugged mountain landforms, the adjacent valleys contain deep accumulations of alluvial and colluvial sediments. These sediments form the Nanyuki Formation, which extends south of the Equator. The Nanyuki Formation consists of gravels, sands, silts, and clays interlayered with colluvial soils and pebbly muds and with numerous tuff horizons. The composition of the component clasts of the epicalstic part of the formation indicates that the precise age of the formation is probably Pleistocene (Charsley 1989:165-166, 172).

In terms of soil, brown loam soil derived from volcanic ash of the underlying Miocene to Pleistocene lava sheets is predominant in the western portion of Laikipia. In the east, red sandy loams derived from volcanic rocks of recent origin appear (Berger 1971:12, 14).

Archaeological research of the Mukogodo Hills (Figure 5) has been underway since 1992 and has continued to the present time with excavations at the Shurmai (GnJm 1) and Kakwa Lelash (GnJm 2), two deeply-stratified dry rockshelters. Kakwa Lelash is located on the foot of the Mukogodo Hills. Although this site belongs administratively to Isiolo District in Eastern Province, the site is located close to the northeast border of Laikipia District. As a result, ecological features of the site are the same as those of Shurmai in Laikipia District.

Ecological Features in North-Central Kenya

In the following section, I describe climate, vegetation, and land use in Kenya in order to understand the ecological features of our study area and the two archaeological sites, Shurmai (GnJm 1) and Kakwa Lelash (GnJm 2).

Climate

In general, three monsoon systems are predominant in the atmospheric circulation over East Africa. These are the dry northeast monsoon, the divergent southeast monsoon and a moist monsoon from the equatorial Atlantic. The pattern of these three monsoon systems brings little rainfall in East Africa. The region north of the equator, including the Mukogodo Hills, is characterized by dry winters between June and August. This is because the tropical rainfall follows the sun southward from north and surface wind from an adjoining dry continental mass reaches this part of Africa. In addition, a dry northeast monsoon prevails during the winter. During the summer the region is affected by cyclonic conditions caused by subsidence of a zonal airflow which comes from very hot Indonesian Archipelago. These cyclonic conditions, called Walker

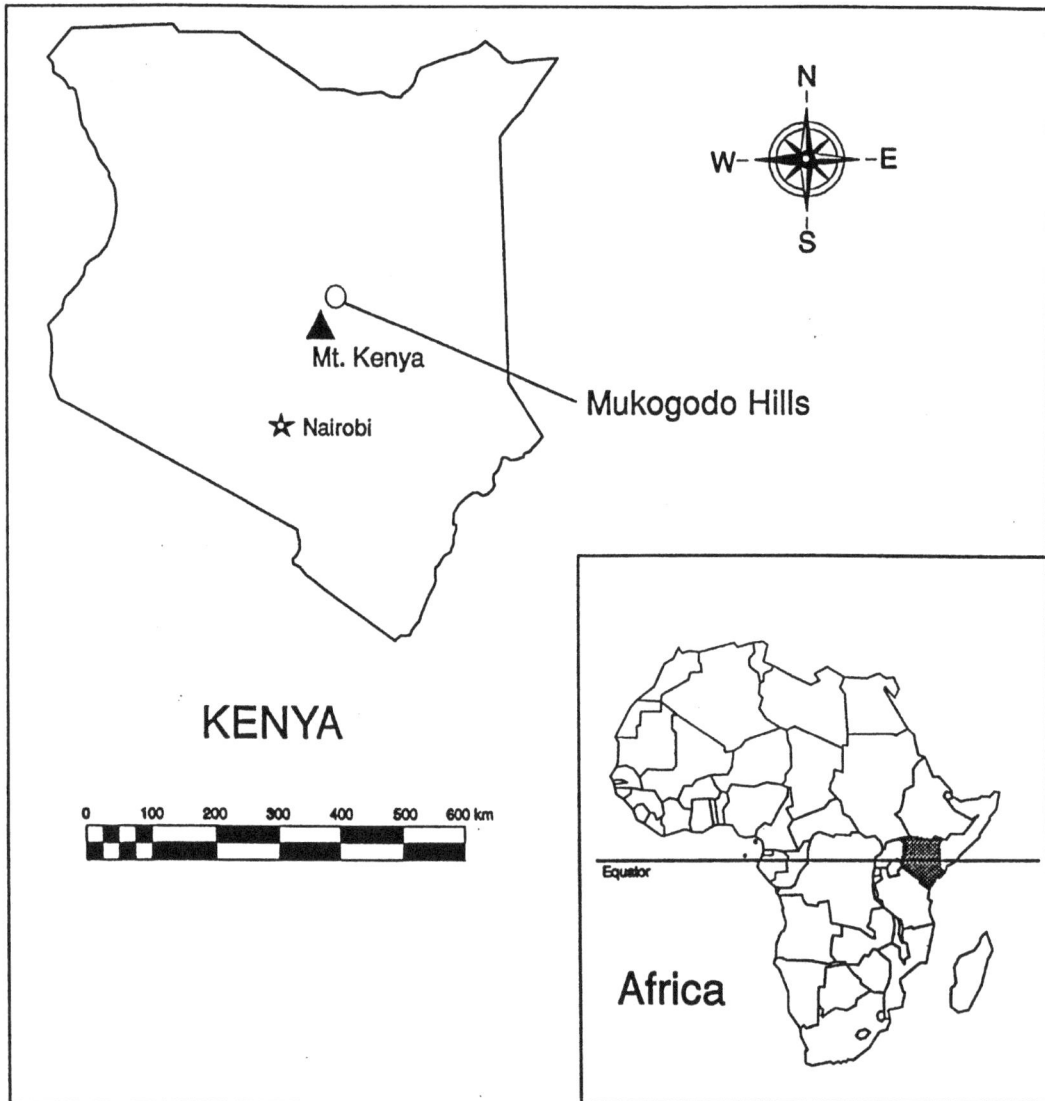

Figure 5. The Location of Mukogodo Hills Region of Laikipia District.

circulation, with the Tropical Easterly Jet prevent high-sun rainfall reaching the region north of the equator in East Africa. However, the region south of the equator in East Africa receives limited rainfall from the Southeast monsoon. In addition, the moisture-laden Atlantic monsoon, which crosses the Congo basin, brings considerable rainfall to the elevated parts. The mountains of Ethiopia, those along the western Rift Valley and the isolated high mountains, including Mount Kenya, Mount Elgon, and Mount Kilimanjaro, on the East African plateau, benefit from this monsoon. Therefore, the divergent southeast monsoon and the moist Atlantic monsoon are the main sources of rainfall for the region north of Equator in East Africa (Coetzee and Bakker 1989:190-191).

The climate of Kenya is typical of the tropics. There are large diurnal temperature variations but small variations in monthly means throughout the year. The seasonal moisture regime of Kenya has two rainy seasons, from March to May and from October to December, following both equinoxes (Schmitt 1992:109).

Within this tropical regional climatic setting, the local climate of the Kenya Highlands is dominated by the effects of an altitude rather than temperature. Rainfall in the area generally increases with altitudes up to 2,450 meters; above this elevation, rainfall decreases. For instance, the highland areas to the east and west of the Rift Valley receive high rainfall totals ranging from 1,000 to 2,000 millimeters. In contrast, the intermediate and low altitude areas including the Laikipia Plateau, the Machakos-Thika districts, and the Rift floor receive less than 900 millimeters. The Highlands area receives between 750 to 2,000 millimeters of rainfall annually. However, compared with areas to the west of the Rift Valley, areas to the east of the Rift Valley, with exceptions only on Mount Kenya and the Aberdares and their slopes have an extremely unreliable pattern of rainfall (Odingo 1971a:12-24).

Sparse climatic data based on mean annual rainfall at Rumuruti (1929-1969) and Don Dol recording stations (1964-1984) indicate that the Laikipia region annually receives rainfall less than 750 millimeters and more than 530 millimeters (Odingo 1971a:Table 2.6; Cronk 1989:Table 1). Brown (1963:4) suggests that lands receiving less than 762 mm of rainfall should be regarded as semi-arid. Ominde (1971:Figure 13.1, 146) suggests that the lands with less than 500 millimeters of rainfall or annual potential evaporation from open water exceeding 2,600 millimeters falls in an arid climate. Taking these two definitions into account, it may be said that the Laikipia District has a semi-arid climate.

There is correlation between altitude and temperature in the Highlands. As altitude increases, mean annual maximum and minimum temperatures decrease. Thus, lower temperatures are associated with higher altitudes. The highest temperatures, in general, occur between January and March, the lowest between June and August. The areas located in the altitude below 2000 meters yield 25 to 27 degrees centigrade of mean annual maximum temperatures and eight to 14 degrees centigrade mean annual minimum temperatures. The area located in the altitude above 2000 meters yield 18 to 23 degrees centigrade mean annual maximum temperatures and eight to 10 degrees centigrade mean annual minimum temperatures. The mean annual maximum temperature of the Mukogodo Hills is 25.7 degrees centigrade in the highlands, while the lowland plains 32.4 degrees centigrade (Cronk 1989:54). The difference between day and night temperatures is apparent because of the elevation and proximity to the equator (Odingo 1971a:12-24).

Biomass

Although there are various ways of classifying vegetation, vegetational physiognomy is considered being very responsive to immediate environmental conditions and physiognomic attributes are important when large-scale classifications of vegetation are required. Physiognomic classification is based on the superficially similar features, called phenotype. Despite species' different histories of independent evolution, the habitants of similar environment resemble one another (Ricklefs 1993:486). Based on physiognomy, Africa as a whole consists of six main vegetation types: lowland tropical forest, montane tropical forest, most savanna, dry savanna, semi-desert and desert, and sclerophyllous scrub (Hamilton 1982:16). According to Hamilton (1982:16-17, Figure 7), lowland tropical forest is characterized by dense woody vegetation which several tree strata and the absence of narrow-leaved grasses in the herbaceous stratum. This type of vegetation occurs within the well-watered tropics at comparatively below 1500-2000 meters. Montane tropical forests appear

similar to lowland tropical forest, but have less stature and fewer tree strata. This type occurs at higher altitudes. Moist savanna is characterized by a single stratum of densely spaced trees with relatively large leaves or leaflets and a herbaceous stratum of tall, narrow-leaved grasses. This type appears in areas that receive less effective precipitation than the forest zones. Dry savanna shows smaller and more widely spaced trees or shrubs, and a lower grass stratum than any other types of vegetation. Semi-desert and desert are characterized by little vegetation cover and are restricted to the driest regions of the continent. Sclerophyllous scrub appears in areas of Mediterranean climate at the northern and southern extremities of the continent. Only this last type of vegetation does not occur in Kenya.

The areas bordering the Kenyan Highlands north and south, and some areas within the arid parts of the country to the east and south are included in a semiarid climate zone. These areas have a dry savanna type of vegetation and are characterized by land of marginal agricultural potential with a natural vegetation cover of dry woodland and savanna formation, often *Acacia-Themeda* association or equivalent bushland (Ominde 1971:151).

Laikipia District has a semi-arid climate and lies within the dry savanna zone of the *Acacia-Themeda* association. The plateau region of Laikipia is characterized by medium to dense growth of mid-grasses interspersed with woody shrubs and trees. The low-lying region of Laikipia is composed primarily of thorny trees and shrubs with variety of grasses. However, the whole area can be divided into three vegetation types: riverine, plateau, and black cotton vegetations (Berger 1971:16-22). The following sections are a summary of Berger.

The riverine vegetation appears along both the Ewaso Narok and Ewaso Ngiro Rivers. Fever trees (*Acacia xanthophloea*) are abundant along the riverbanks. Farther away from the rivers Wait-a-bit thorn (*Acacia brevispica* and *Harpagophytum procumbens*) and Sudan Gum Arabic (*Acacua senegal*) are so dense as to be almost impenetrable. A few scattered spurges (*Euphorbia candelabrum*, *Euphorbia grandicornis*, and *Euphorbia robecchii*) occur on dry and stony sites. Cacti of two kinds, *Opuntia ficusindica* and imperfectly known *Opuntia* species, are present. A large number of grass species occur in this region. They include common star grass (*Cynodon dactylon*), bamboo grass (*Pennisetum mezianum*), Naivasha star grass (*Cynodon plectostachyum*), and Masai grass (*Pennisetum stramineum*).

Plateau vegetation type appears west of the escarpment. The most dominant grass species is red oat grass (*Themeda triandra*). Many other grasses are also present. They are bamboo grass, mountain needlegrass (*Aristida adoensis*), Masai love grass (*Eragrostis superba*), and so on. Trees are widely scattered with whistling thorn (*Acacia drepanolobium*). Nile acacia (*Acacia nilotica*) is most abundant. Black Cotton vegetation occurs on poorly drained areas of the plateau and is called the *Acacia-Pennisetum* woodland. The woody vegetation is composed of whistling thorn and white-galled acacia (*Acacia seyal*).

Today the Mukogodo hills support a unique dry tropical forest assemblage consisting largely of wild olive (*Olea africana*), cedar (*Juniperus procera*), and candelabra tree (*Euphorbia sp.*). In addition, various thorny *Acacia* species and a host of other taxa occur (Cronk 1989; Dickson, Kuehn, and Cronk 1995: 1). Rainfall, temperature, and elevation largely affect the distribution of arboreal vegetation: cedar and olive, in general, appear in highland, while acacia, candelabra and other various dry land succulents are mainly distributed on the lowlands as one descent from the forest margins to the north and east (Dickson and Kuehn 1997:3).

Animal life in the Mukogodo Hills is also varied. The East African leopard (*Panthera pardus*) is a major predator in the forest, while the lion (*Panthera leo*) appears on the open savanna country on the forest margins. Elephants (*Loxodonta africana*) and black rhinoceroses (*Diceros bicornis*) were common in the forest. Both large and small ungulates are present in the forest. Deer, Cape buffalo, antelope, and wild asses also appear. Most common animals in the savanna are the rabbit-sized dikdik (*Madoqua*). Bush pigs, forest pigs and wart hogs are present as well as such primates as baboons and colobus and numerous other species of monkeys appear. On the surrounding savanna, giraffes, zebras and oryx are abundant. The lowly rock hyrax or dassie (*Heterohyrax brucei*) small, furry rodent-like creatures which appear distantly related to the elephant are present as well (Dickson and Kuehn 1997:3).

Land Use

In Kenya land has always been at the center of politics, economics, and social life. Because of the lack of significant minerals or of a significant mining industry, most of the economic development of the country has been concentrated on agriculture (Odingo 1971b:162). Prior to European settlement, there were various types of land use. These patterns are closely related to the cultural sequence in the region, and will be discussed in the following section.

According to Ominde (1971:153-154), the present pattern of land use is closely associated with three main stages in the evolution of the economy. The first phase is characterized by the establishment of European commercial farming in the Highlands and other parts of the country. This first phase led to a dual approach to land utilization, and increased the gap between those parts of lands in the former European settled areas and the remaining African areas. This first phase will be discussed later in terms of the land use pattern of the Highlands. The second phase is the period of the African Land Development Organization and the Ten-Year Development

Program 1945-1955, and its extension from 1954 in the form of the Swynnerton Plan. That is, the colonial policy of the post-war period. The third phase is a post-independence phase of planned national development. This phase emphasizes development of high-potential land rather than utilization of uninhabited area. As most of the arid and semiarid land has potential for animal domestication, these areas begin to be under the developmental scheme.

Under the traditional system, agriculture was the least important subsistence mode in the arid and semiarid areas and is severely restricted by erratic and low rainfall. The new approach to agricultural problems of these areas has involved maximum use of limited agricultural potential through selection of suitable crops and through intensification of production in ecologically suitable areas. In this course, land policy on agriculture emphasizes three activities. The first is commercial farming. From the beginning of commercial farming in Kenya, sisal has been a vital crop element in the semiarid areas. The second is irrigation development. This form of agriculture is practiced in the more humid areas with better communications. The national significance of irrigation lies not merely in providing an alternative form of land use but in meeting local and national food requirements, as well as providing an additional source of revenue. Under this situation, four areas are considered as possibilities of irrigation. They include the Lake Turkana Basin, the Lake Baringo Basin, Ewaso Ngiro Basin, and Tana Basin. The last agricultural activity is development of the pastoral base of the economy. The new emphasis on the pastoral economy takes into account the natural suitability of the area, the important contribution of the arid and semiarid areas to the national economy, and findings on well-grounded methods of balanced use of the natural potential (Ominde 1971:154-157).

However, only about one third of the country has high potential for agricultural development. Most of the area is in the Highlands. Two sides to the agricultural system in the Highlands had developed before independence. One was large-scale plantation farming of coffee and sisal, the other is a comparatively small-scale arable farming based on maize, wheat, pyrethrum, and some coffee. In mixed dry farming, wheat and maize clearly emerged as the predominant cereals and main cash crops. Areas where mixed farming was predominant in the Highlands are restricted by ecological as well as economic factors. Most of the mixed farming areas are located in the arable and better-watered parts of the Trans-Nzoia, Uasin Gishu, Sotik, Londiani, Molo and Mau Narok. This dichotomy between plantation and mixed farming continues in the agricultural system into the present (Odingo 1971a:182-183).

In addition to these two agricultural systems, dairying has emerged as the stable and most established of the livestock industries. In the livestock industries, goats, sheep, cattle, and camels are major species being raised (Dickson 1997, personal communication). The delimitation of the livestock farming regions is based on the agricultural values. That is, the areas where pastoralism is dominant are isolated from arable land although specialization of dairying, beef farming, or even woolen sheep farming is present within the mixed farming areas. The ranching and dairy ranching areas present the best example of how ecological conditions affected farming in the Highlands. The main ranching districts include Naivasha, Laikipia, Mount Kenya and parts of the Machakos and Thika districts. The difficult conditions in these areas, in particular the low rainfall, make livestock farm possible only on a ranching basis. Crop farming within these areas was clearly a risky affair because of the lack of rainfall and its unreliability (Flury 1988:267). Frequent crop failures encouraged the farmers within these areas to specialize in beef ranching with the occasional Merino sheep ranch and dairy ranching in the areas where it paid to produce milk even under difficult conditions (Odingo 1971a:164-170, 182-185).

Cultural Sequence and Cultural Contact of the Mukogodo Hills Region

Cultural sequence and cultural contact in East Africa are not clearly known. Many scholars (Phillipson 1993, Cole 1963) classify African people and its cultures on the basis of linguistics. Reconstruction of the cultural change in Kenya is risky attempt because there is no adequate documentary evidence. However, in order to understand the composition of the present population in Kenya and the study area, reconstructing cultural sequence is useful. In this section I will describe cultural sequence in the study area based upon linguistics and archaeological evidence.

The Mukogodo are a combination of peoples who has different origins. Based upon linguistic evidence, Mukogodo oral traditions, and the oral traditions of their neighbors, the Mukogodo consist of three or four different sources (Cronk 1989:38). They are the Eastern Cushitic language family, a Southern Nilotic language family, and the Khoisan language family. The Mukogodo language, Yaaku, belongs to the Eastern Cushitic sub-branch of the Afro-Asiatic language family (Greenberg 1963). In the early second millennium B.C. the Yaakuan language group became distinct from the related Werizoid language in and around the area that is now southern Ethiopia, and by around 1000 B.C. Yaakuan speech was widely spread over a most part of East Africa, including northern Kenya where they were the first Eastern Cushitic settlers (Ehret 1976:93). Although these original Yaakuan-speakers were probably herders like other Eastern Cushitic speakers, later Yaaku became a language of hunters (Cronk 1989:39). The origin of the word AYaaku means hunters. The word AYaaku was borrowed from a southern Nilotic language sometime around the end of the first or the beginning of the second millennium A.D. Although it is not clear whether the hunter-gatherers who spoke Yaaku were descendants of the pastoralist Yaakuan-speakers or of

indigenous foragers who borrowed the language during the period of Yaakuan dominance, it is believed that the original language of these hunter-gatherers may have been related to Sandawe, a member of the Khoisan language family. Today Sandawe language which is a distant relative of the language of the !Kung San of Botswana and Namibia and of the Hadza of Tanzania is spoken in Tanzania (Ehret 1974:33). That is, of three linguistic sources, eastern Cushitic is the latest influence on the Mukogodo. Somali, Galla, Afar, Sidamo, and Beja languages are also included in Cushitic (Phillipson 1996:6).

Today Cushitized Nilotes are found in three main clusters. They are the Karamojong, the Nandi, and the Masai. The Turkana and Topotha in the extreme northwestern parts of Kenya's drylands represent the Karamojong. The Nandi cluster is represented by Elgeyo, Tugen, Suk, and Marakwet. South part of the region is occupied by remnants of the Masai. Groups such as the Samburu, Elmolo, Laikipia, Mukogodo, and Njamus are included in the Masai cluster. The most recent movement of the Cushitic people in the region is indicated by the presence of the Galla and the Somali populations, which are a pastoral people. The presence of the Galla people in drylands of Northern Kenya is largely due to pressure exerted against them by Cushitic Muslim neighbors. The contemporary presence of the Somali population strengthen the pastoral culture complex of the arid and semi-arid region because these people make their living primarily from camels, sheep, goats and cattle (Ominde 1971:152-153).

The Mukogodo Hills are the traditional home of the modern Mukogodo Masai people. Excavation of the stratified Shurmai Rockshelter in the Mukogodo Hills in 1993 and 1994 revealed a sequence of human occupation extending from the present back to the Later and Middle Stone ages (Dickson 1995; Dickson and Kuehn 1995). Based on archaeological work conducted by Dickson over the last four years in the Mukogodo Hills regions, a foraging way of life appeared sometime after about A.D. 850. Although there is not clear evidence of a preceding Pastoral Neolithic presence in the study area, there is a possibility that the extensive valley side erosion that forms the massive slopewash aprons and alluvial fans is caused by overgrazing that began with the arrival of herding peoples during the Neolithic period (Dickson, Kuehn, and Cronk 1995:8).

According to ethnography (Cronk 1989), the Mukogodo people began a remarkable social and cultural transformation in the 1920's. Since 1936, they abandoned their foraging way of life and adopted the pastoral culture of their Masai neighbors. The total cultural transformation of the Mukogodo into Masai has taken less than two generations. Today the Mukogodo abandon their deep forest rockshelter sites and occupy open settlement locations near good pasture.

In summary, the Mukogodo Hills region dominates the semiarid landscape and supports a dry tropical forest assemblage composed mainly of wild olive and cedar. A series of river valleys dissects the range. The river valley supports *Acacia-Themeda* association. Based on geological surveys, the basement rocks of the eastern portion of the Laikipia District indicate that they are among the oldest on the African continent. On the other hand, the sedimentary deposits that fill the valleys show very recent origin. This configuration of landform and geology in the district has important implications for the prehistory of human occupation there.

Archaeological Research of the Mukogodo Hills Region

Multi-disciplinary research of the Mukogodo Hills and environs in the Laikipia District of Kenya has been underway for five years. This research was initiated as four-week site reconnaissance during the dry season of 1992 and has continued to the present time with excavation work and further site survey. During the site reconnaissance full twenty-three sites had been located and twenty-one of them carefully sketch-mapped. During the 1994 dry season, two additional rockshelters were included in this corpus. As a result, a total of twenty-five sites was surveyed (Figure 6).

All of the cave and rockshelter sites included in the corpus yield a sign of human occupation and a number of them have deposits that indicate a high degree of site resolution. As the interiors of many of them are shielded from direct rainfall, most of the cave and rockshelter sites are characterized by low humidity and permanently dry floor deposits. Due to the absence of water, environmental transformation by organic decomposition and dissolution is minimal and such materials as plant and animal remains, fecal material, human remains, hyrax middens, and organic artifacts are well preserved. The deposits in these caves and rockshelters have not been disturbed and their contents reclaimed by illicit diggers. Unlike caves and rockshelters in many parts of the world, the archaeological sites in the Mukogodo Hills are largely or completely intact. The preservation of these sites is due both to their remote location and to the low public interest in pothunting and grave robbing among Kenyans. The site reconnaissance of 1992 laid the groundwork for the field research in the Mukogodo Hills, test excavations in the Shurmai Rockshelter (GnJm 1) and Kakwa Lelash Rockshelter (GnJm 2). The following description is the result of archaeological excavation by Dickson (Dickson 1993a; Dickson and Kuehn 1995, 1997; Dickson, Kuehn, and Cronk 1995; Kuehn and Dickson 1997).

Test Excavations in the Shurmai Rockshelter (GnJm 1)

The Shurmai Rockshelter (GnJm 1) is located at E 37° 12.911' and N 0 ° 30.5" at an elevation of 1280 meters above the sea level in the Mukogodo Hills region. The rockshelter sites are composed of four individual cavities that run the southwestern cliff face of the Shordika Hill which is characterized by a prominent granite gneiss inselberg that locates the north of the Shurmai valley.

The four individual cavities of the rockshelter site run from east to west along the Shordika cliff face. Of these four cavities, the cavity called A is the largest and most westerly. Although four cavities are studied, the cavity A is the most interest. Therefore, in the present study, the cavity A represents the Shurmai Rockshelter.

The Shurmai Rockshelter is a long, lozenge-shaped cavity of approximately 37 meters from east to west and about 15 meters from the drip line at its lip to its back wall. From the floor of the shelter to its overhanging roof is 15 meters. The granitic gneiss bedrock of the cavity is exposed at the lip of the shelter. The floor of the shelter inside the drip line is littered with rock fragments produced by roof and wall collapse.

Over a four-week period during the dry season of 1993, preliminary test excavations were conducted in the Shurmai Rockshelter, which was the most promising site in the corpus of twenty-three sites assembled during the first phase of research. A detailed contour map of the interior surface of the Shurmai site provided a base for the excavations. The excavation grid was accurately established by measuring slope distances and calculating horizontal distances and measuring angles to the nearest thirty seconds. The map was later drawn using Auto CAD and QUICKSURE mapping programs (Figure 7). The site was precisely located using the Global Position System (GPS). All maps and field and laboratory data in the high precision Geographic Information System (GIS) for quick retrieval were stored. Artifactual and zooarchaeological materials found on the surface of the site were also systematically collected by grid square. In order to expose a complete stratigraphic and cultural sequence there, a test trench was excavated in the rock shelter. Excavation work at Shurmai continued during the dry season of 1994. During the excavation in 1994 the initial test excavations were expanded and a two-by-two meter *sondage* was excavated to bedrock. During both field seasons, excavations were conducted by natural layers and each unit was described in terms of its granulometry, calcimetry, roundness, concretions, and mineralogy in order to interpret the source, mode of deposition, and degree of post-depositional modification. Samples were extracted for (1) radiocarbon dating, (2) pollen analysis, (3) soil characterization studies, and (4) stable isotope analysis. Data collected from two field seasons excavation revealed that the Shurmai Rockshelter consists of 4 Units (Figure 8).

Of these, Unit 1 has two subunit, 1a and 1b, and Unit 4 has 3 subunit, 4a, 4b, and 4c. However, Unit 1 (depth of 113-200 cm) does not yield any cultural material. Unit 2 (depth of 92-113 cm) characterized by the oldest cultural horizon

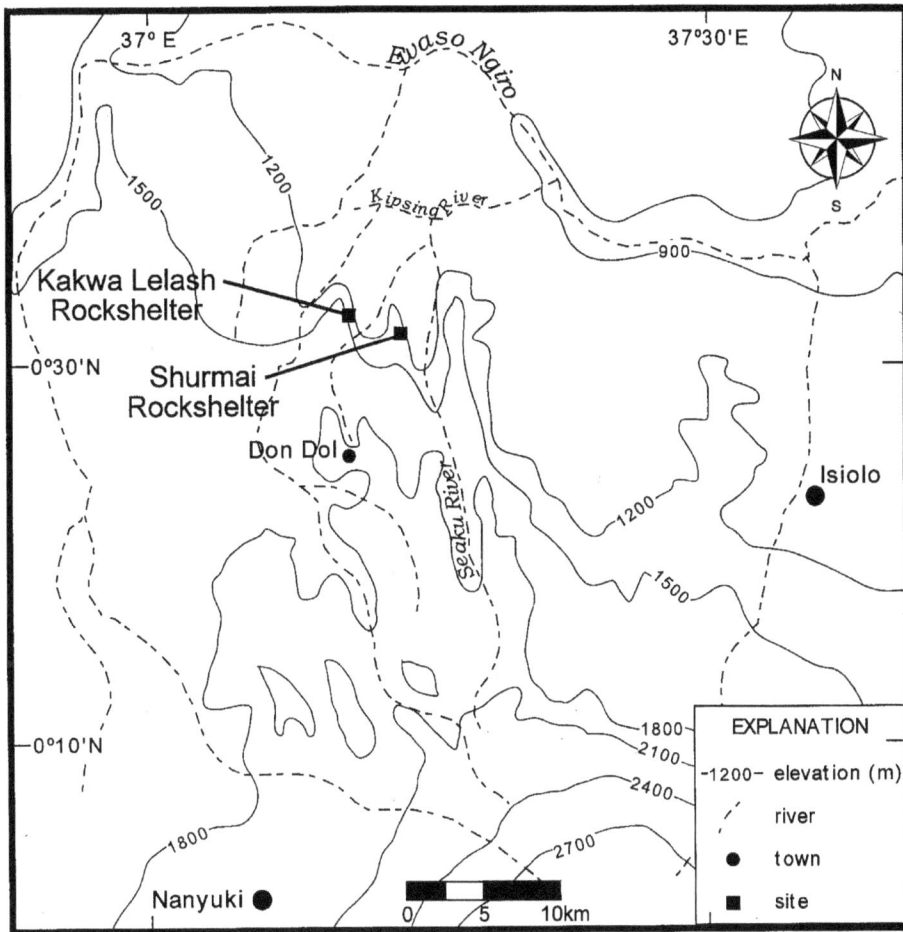

Figure 6. Location of Shurmai Rockshelter (GnJm 1) and Kakwa Lelash Rockshelter (GnJm 2) in the Mukogodo Hills region.

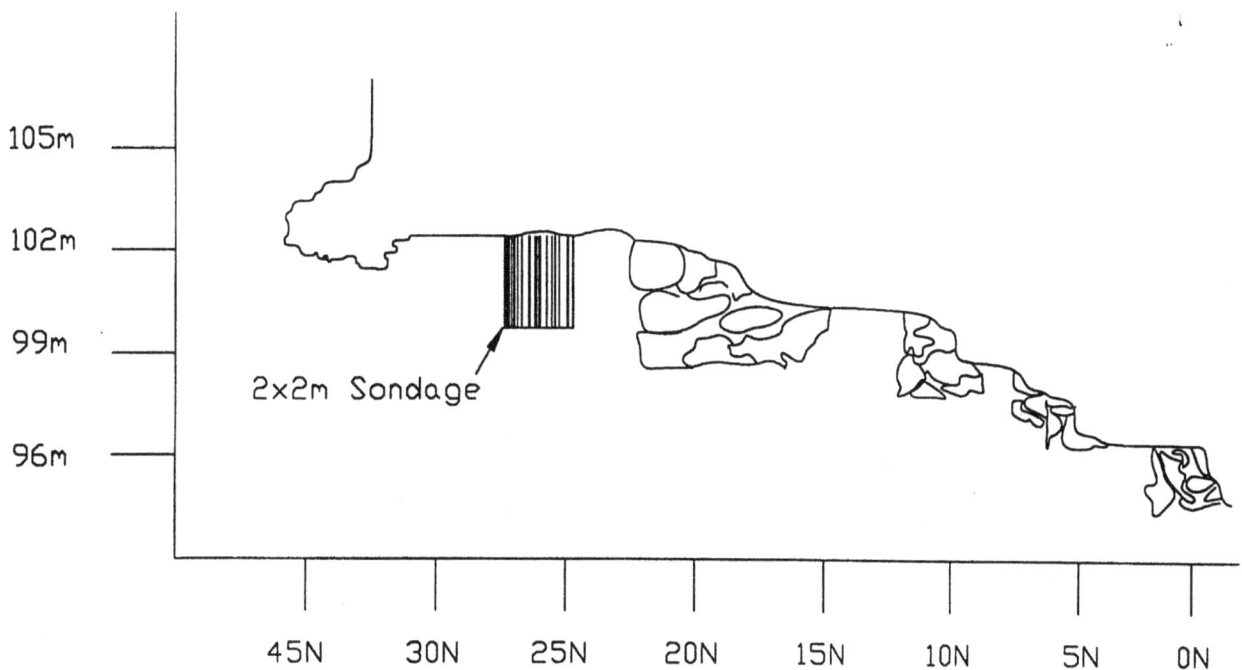

Figure 7. Topography of Shurmai Rockshelter (GnJm 1).

GnJm1 SHURMAI ROCK SHELTER
East wall profile

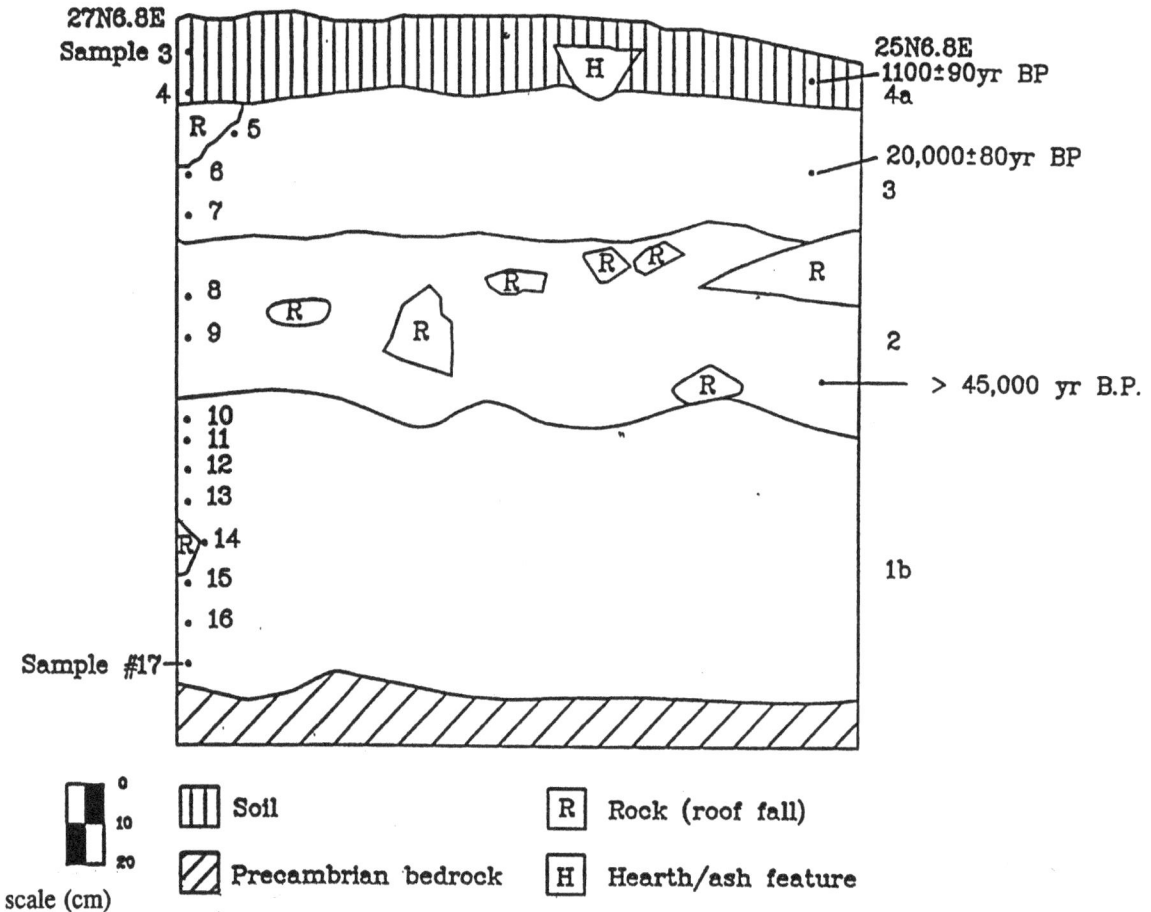

27N6.8E
Sample 3
4
R .5
. 6
. 7
. 8
. 9
R
R
R R
R
R
R
. 10
. 11
. 12
. 13
R .14
. 15
. 16
Sample #17

25N6.8E
1100±90yr BP
4a
20,000±80yr BP
3
2
> 45,000 yr B.P.
1b

H

scale (cm)
0
10
20

Soil

Precambrian bedrock

R Rock (roof fall)

H Hearth/ash feature

Figure 8. Stratigraphy of Shurmai Rockshelter (GnJm 1).

in the sondage. It produces a high density of lithic materials and small amounts of fossilized bone. Infrared stimulated luminescence (IRSL) analysis measurements of feldspar grains using the single aliquot method yielded an age of 45,211 ± 5356 yr BP. Unit 3 also contains dense accumulation of faunal remains and lithic material. The date of Unit 3 obtained from a sample of organic material recovered from Unit 3 (depth of 65-92 cm) is an accelerator mass spectrometry radiocarbon age of 20,000 ± 80 yr BP (Beta 85593). Unit 4 (surface to 65 cm) subdivided into three subunit is characterized by abundant cultural remains in the form of flaking debris, bone, pottery, and historic and modern material at the surface. A sample of charred organic material recovered from the subunit 4a (28 centimeter thickness) yielded a standard radiocarbon age of 1100 ± 90 yr BP (Beta 85544).

Test Excavations in Kakwa Lelash Rockshelter (GnJm 2)

When bedrock was reached in the Shurmai Rockshelter without encountering any evidence of a hominid presence beneath the Middle Stone Age occupation layer, it was concluded that sequence-building goals would be better served by shifting attention to another rockshelter, Kakwa Lelash Rockshelter (GnJm 2), that might contain evidence of earlier occupation.

This site is located near the top of a large Precambrian metamorphic erosional remnant or inselberg that rises some 100 meters above the surrounding valley floor. The shelter itself appears to have formed by the erosion of exposed bedrock by running water. The resultant cavity is now almost completely filled with alluvial and aeolian sediments together with roof fall and other forms of colluvium. Both historic and prehistoric debris present on the surface of the shelter indicates a long occupation history. Over a six-week period in the dry season in 1995,

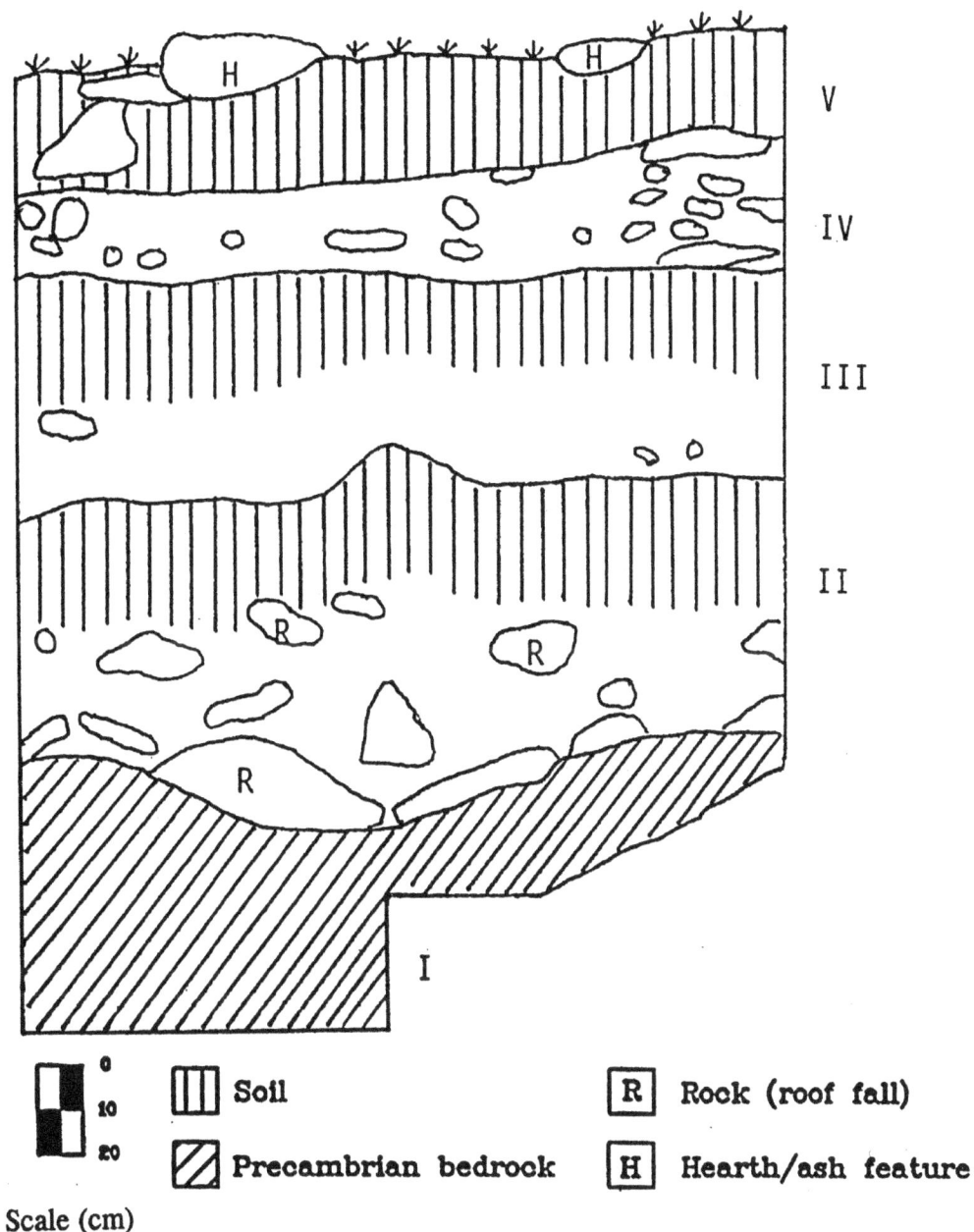

Figure 9. Stratigraphy of Kakwa Lelash Rockshelter (GnJm 2).

a deep two-by-two meter *sondage* in the deposits at Kakwa Lelash Rockshelter was excavated to bedrock at a depth of slightly less than three meters. Kakwa Lelash Rockshelter consists of 5 units: Unit 1 (depth of 166-230 cm), Unit 2 (depth of 113-166 cm), Unit 3 (depth of 51-113 cm), Unit 4 (27-51 cm), and Unit 5 (depth of 0-28 cm) (Figure 9). Unlike Shurmai Rockshelter, lithic material excavated from Kakwa Lelash Rockshelter is almost evenly distributed through the units. However, the most abundant lithic material was recovered from Unit 3. Both Unit 1 and Unit 5 produced less lithic material than other units. Analysis of the stratigraphic units of Kakwa Lelash Rockshelter is still underway. As was the procedure in the Shurmai excavations, the stratigraphy exposed in these test

excavations was closely examined and appropriately sampled and all lithics of a putative Middle Stone Age provenience were piece plotted. Regrettably, aside from the occasional object of possible MSA affiliation, little evidence of an occupation earlier than the Later Stone Age was found in Kakwa Lelash rockshelter. At this time only preliminary results of this seasons fieldwork are available and a full cross dating of the Kakwa Lelash Rockshelter with the Shurmai Rockshelter is currently being undertaken.

The exploratory excavations of the Shurmai and Kakwa Lelash Rockshelter sites have now revealed that the region was occupied at least during the Middle Stone Age.

Preliminary analysis of a surface lithic scatter discovered on a rock terrace of the Tol River during the 1995 season suggests that hominid presence in this region may extend back to the Early Stone Age. The systemic analysis and comparison of the lithic materials recovered in the Shurmai and Kakwa Lelash rockshelter sites, which I propose to undertake, will be a key part of this study undertaking and will lay the foundation for further research in the prehistory of this rich and still largely an unknown region.

Raw Material Source Area

As mentioned above, the Mukogodo Hills region is located heavily forested range of low mountains and inselbergs. On this region, Shurmai Rockshelter (GnJm1) lies on the eastern rim of the Laikipia Plateau near the boundary between Rift Valley and Eastern provinces of north central Kenya. Kakwa Lelash Rockshelter (GnJm 2) is also located on the extreme eastern rim of the Laikipia Plateau which is administratively belongs to Isiolo District of Eastern Province. Therefore, these two archaeological sites lie on the same geological setting. The rock outcrops of the Mukogodo Hills are characterized by Precambrian to Quaternary age. As the region is associated with East African Rift Valley, which resulted from serious volcanism, both igneous and metamorphic rock types are dominant. As of the present time, no study of raw material source area has been undertaken. Therefore, in this section, I describe rock source areas on the basis of rock types that are found in the present study. The characteristic or definition of each rock type will be described in the Chapter V.

Volcanic Sources

Volcanic rocks found in the present study include both igneous rocks, such as obsidian, basalt, rhyolite, tuff, and metamorphic rocks, including quartzite, gneiss, and schist. Of these, basalts and obsidians are dominant. The rest of volcanic rocks are represented by several samples.

Obsidian. Dark-colored volcanic glass, obsidian, is favored by both prehistoric and historic knappers. In Kenya, at least 54 geographically discrete obsidian source localities have been located and collected. The majority of obsidian bearing localities are centered around the Lake Naivasha basin, Mount Eburru, the eastern highlands flanking the central Rift Valley. Relatively minor sources of obsidian is also found the northern portions of Kenya, east of Lake Turkana, in the southern end of the Suguta valley, on Mount Kenya, in the southern Kenyan rift zone (Merrick and Brown 1984:129-135).

The nearest known obsidian source area is Mount Kenya, which is located approximately 70 kilometers from the two archaeological sites. Mount Eburru and the source areas around the Lake Naivasha basin are located approximately 200 kilometers from the present study area. Mount Kenya is located southeast and both MountEburru and the Lake Naivasha southwest of the present study area.

Basalt. In geomorphological terms, both the Shurmai (GnJm 1) and Kakwa Lelash (GnJm 2) rockshelters are located on inselbergs. In general, inselbergs, which are residual bedrock knobs, stand above the general level of the pediment. A pediment is an erosional surface that abuts against and slopes away from a mountain front or escarpment. The erosional surfaces of the pediment are frequently cut on the same rocks that make up the mountain. Pediments are best developed and preserved on bedrock particularly resistant types, including granite or related crystalline (Ritter 1986:286, 288). In the Mukogodo Hills region, the inselbergs containing two archaeological sites (GnJm 1 and GnJm 2) are dissected by several rivers and streams that are diverged from Kipsing River. Of these several rivers, Tol River passes near Kakwa Lelash Rockshelter and Seaku River near Shurmai Rockshelter. In the research area, the nearest known basalt source area can be found in the place where Tol River and Seaku River start diverging from the Kipsing River. This place is located 7 kilometers from Kakwa Lelash site and 10 kilometers from Shurmai sites. However, based on geomorphic study, there is a possibility to find other close basalt source areas, which are located six kilometers from both sites. Today this place does not show basalt outcrops, and so the basalt of this area might be buried (Kuehn 1997, personal communication). The nearest known basalt place lies northeast of the two archaeological sites.

Rhyolite. Like obsidian, rhyolite is a granitic rock composed of potassium feldspar and quartz. Among granitic rocks, including granite, obsidian, and pumice, rhyolite is rarely found. In the present study area, granitic rock source areas have not been found within 50 kilometers from the archaeological sites.

Gneiss and Schist. Only few archaeological samples of these materials were discovered from the present study area not because of the lack of the raw material source itself, but because of the poor quality of the flaking these materials. These two raw materials are found at the study area.

Quartzite. Few archaeological samples of quartzite were found. The quartzite source area is not found within 20 kilometers from the study area.

Non Volcanic Sources

Predominant non-volcanic raw materials found in the present study area are quartz and chert.

Quartz. As mentioned above two archaeological sites, Shurmai (GnJm1) and Kakwa Lelash (GnJm 2) are located on the inselbergs. In general, most inselbergs represent areas of rocks that are characterized by very resistant to weathering and erosion (Ritter 1986: 288). The present study areas on the Mukogodo Hills region contain many quartz source areas.

Chert. Chert deposits are commonly found either as irregularly shaped nodules in limestone or as layers of rock. The silica that comprises most chert nodules is believed to have been deposited directly from water. Particularly beds of chert are thought to have originated mainly as biochemical sediment. Some bedded chert occurs in association with lava flows and layers of volcanic ash. In this case, silica is derived not from biochemical sources, but from the decomposition of the volcanic ash (Tarbuck and Lutgens 1993:145-146). Within 10 kilometers distance from the present study area, chert source area has not known. However, as the Laikipia

Plateau is formed by lava flow and other volcanism, it is possible to find chert source area within 50 kilometers distance from the present study area.

In summary, the raw materials found in local sites are generally igneous and metamorphic rocks. Raw material source areas for basalt, quartzite, gneiss, and schist are located approximately within 10 kilometers from the site. However, the nearest known source areas for obsidian is located approximately 70 kilometers from the sites. Of non-volcanic rocks, quartz is mainly found within 10 kilometers from the sites. Although we did not found source areas for chert within 10 kilometers from the sites, chert source areas probably can be found with 50 kilometers from the sites (Table 1).

Table 1. Raw Material Source Areas

Volcanic	Rock Type	Distance from the Sites	Direction
	Basalt	< 10 kilometers	Northeast
	Quartzite	< 10 kilometers	Site
	Gneiss	< 10 kilometers	Site
	Schist	< 10 kilometers	Site
	Obsidian	> 50 kilometers	Southeast and Southwest
Non volcanic	Quartz	< 10 kilometers	Site
	Chert	< 50 kilometers (?)	West (?)

Chapter III
Lithic Technology

One of the major goals in this dissertation is to reconstruct and compare the lithic technology at two rockshelter sites, Shurmai (GnJm 1) and Kakwa Lelash (GnJm 2). In lithic archaeology, the terms "technique" and "technology" convey different meanings. Technique means skills and methods, which are applied to accomplish tasks, while technology consists of technique and the totality of rational methods (Crabtree 1972:2). In other words, technology refers to "an assemblage of human beings and apparatus in structured relationships designed to produce certain specified results" (Winner 1977:75). The technology, thus, involves the totality of the various processes beginning with the procurement or raw material or resource acquisition, manufacture including its preparation, the production of flake blanks, and their final retouching into finished tools, manipulation, and discard or loss (Debénath and Dibble 1994:22; Hayden et al. 1996:37; Newcomer 1975:98).

The Nature of Raw Material

It is necessary to discuss the nature of raw material suitable for flint knapping because application of lithic technique to produce desirable flakes or tools partly depends on types and nature of raw material. The important factors in flint knapping both in the prehistoric time and the present are not only the acquisition of abundant stone sources, but also the procurement of good quality of stone material. Not all stone material is flakable. The capacity of the knapper to shape the tools and functional performance of the tools are influenced by the quality of the raw material. In this section, I will discuss the stone quality and stone materials suitable for the knapping based on archaeological samples that were recovered from the two sites, Shurmai (GnJm 1) and Kakwa Lelash (GnJm 2).

Stone Quality

The first requirement is that the stone materials used in flint knapping must fracture conchoidally. A conchoidal fracture refers to a diagnostic fracture on a plane surface. It resembles half a bivalve shell, indicating definite striking patterns. In conchoidal fracture, the striking area usually at the hinge part of the bivalve shell and the conchoidal fracture appear on the part below that hinges (Crabtree 1972:54). Furthermore, the stone should be homogenous, isotropic, brittle, and elastic. Homogeneous material can be worked with consistency because it is the same throughout and lacks differences in texture, cracks, inclusions, flaws, and irregularities. If the material is not homogeneous, it may not fracture conchoidally since cracks, planes of weakness and other flaws impair the conchoidal fracture process and make a stone break unpredictably or in undesirable direction. The most desirable stones for flaking have amorphous or cryptocrystalline structures. Cryptocrystalline structure refers to the texture of a rock consisting of individual

crystals that are too small to be recognized with the naked eye (Bates and Jackson 1984:120). This means that the rock suitable for flaking should be fine-grained. The homogeneity of the crystalline structure is easily recognized in the texture of a freshly flaked surface. The crystalline structure is also the most important factor in determining how easily a material can be worked. The most homogeneous materials with amorphous crystalline structures are man-made glass and obsidian, a natural black or dark-colored glass produced by volcanic action. These homogeneous materials fracture very easily and produce extremely sharp edges and smooth fracture surfaces. On the other hand, cryptocrystalline materials such as flint and chert are harder than man-made glass or obsidian and therefore more difficult to fracture than obsidian and do not form such sharp edges. In this case, the fracture surfaces are smooth to slightly rough and it is not possible to recognize crystal structure in texture of a rock without aid. The toughest and least amorphous materials, such as dark-colored igneous rock basalt or metamorphic rock quartzite, are hard to work. These materials are characterized by the rough fracture surface with a grainy or sugary texture (Whittaker 1994:12, 68-69). Another important property that is closely related to homogeneity is isotropy. Isotropy refers to the properties that are uniform in all direction. Thus flaking isotropic materials can be predictable.

Elasticity refers to "the property of stone to return to its former state after being depressed by application of force" (Crabtree 1972:60). If the stone that is deformed under some applied force or load returns to its original shape after the load is removed, it is said to be elastic. Brittleness means the ease of breaking the rock without much deformation. In general, a brittle material has no plastic region and continues to behave elastically as the application of force is increased up to the point where it breaks. In contrast, as ductile materials which is characterized by their strong resistance to deformation before fracturing or faulting behave elastically only under relatively small loads, the material deforms permanently or plastically when the applied force is increased beyond some critical yield point (Speth 1972:36). Errett Callahan (1994:11) has proposed a Lithic Grade Scale to classify raw materials. His classification is useful to understand stone quality in flint knapping (Figure 10).

Stone Materials Suitable for Flaking

Before I discuss stone materials that are desirable for flaking, it is necessary to distinguish between the terms, "mineral" and "rock." A mineral is a naturally occurring inorganic solid that possesses an internal structure and a chemical composition. Both an orderly internal structure and a definite chemical composition of mineral determine the properties of rocks. A mineral consists of an ordered array of atoms to form a particular crystalline structure. These ordered arrays of atoms are called crystals. The two most abundant mineral elements of the rocks of the earth's crust are silicon and oxygen. The combination of silicon and oxygen produce the most common mineral group, the silicates. As all silicate minerals have the fundamental

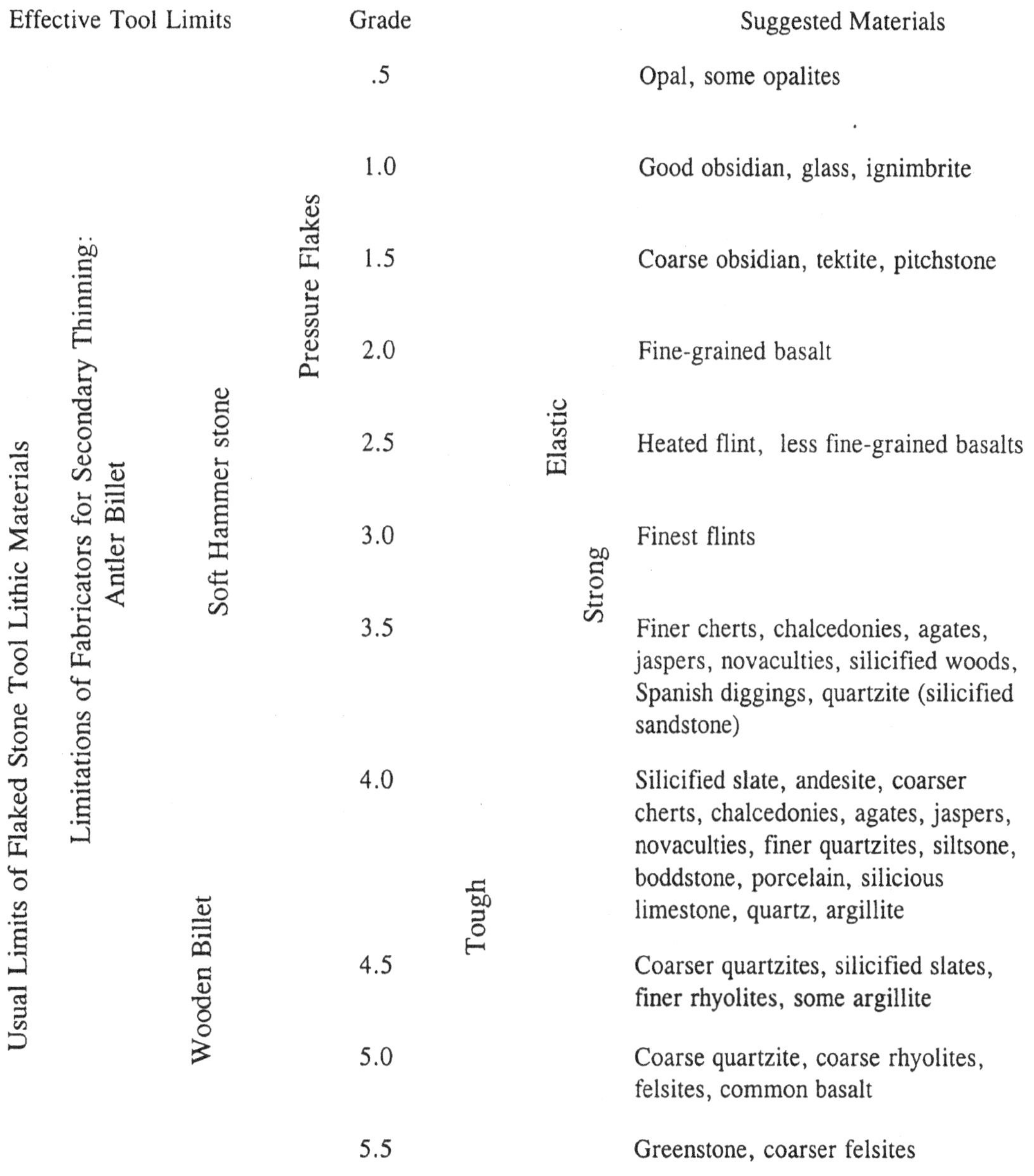
Effective Tool Limits
Grade
Suggested Materials
.5
Opal, some opalites
1.0
Good obsidian, glass, ignimbrite
1.5
Coarse obsidian, tektite, pitchstone
2.0
Fine-grained basalt
2.5
Heated flint, less fine-grained basalts
3.0
Finest flints
3.5
Finer cherts, chalcedonies, agates, jaspers, novaculties, silicified woods, Spanish diggings, quartzite (silicified sandstone)
4.0
Silicified slate, andesite, coarser cherts, chalcedonies, agates, jaspers, novaculties, finer quartzites, siltsone, boddstone, porcelain, silicious limestone, quartz, argillite
4.5
Coarser quartzites, silicified slates, finer rhyolites, some argillite
5.0
Coarse quartzite, coarse rhyolites, felsites, common basalt
5.5
Greenstone, coarser felsites
Usual Limits of Flaked Stone Tool Lithic Materials
Limitations of Fabricators for Secondary Thinning:
Antler Billet
Soft Hammer stone
Wooden Billet
Pressure Flakes
Elastic
Strong
Tough

Figure 10. A Proposed Lithic Grade Scale for Flaked Stone Tool Materials.

building block, the silicon-oxygen tetrahedron, silicate minerals tend to cleave between the silicon-oxygen structures rather than across them.

A rock is a consolidated mixture of one or more minerals. Rocks can be divided into three general types: igneous, sedimentary, and metamorphic. Igneous rocks are formed from the crystallization of magma. Magma is naturally occurring molten rock material found at depth of the earth. Sedimentary rocks are composed of the weathered products of pre-existing rocks that have been transported and deposited. Metamorphic rocks result from the alteration of pre-existing rock deep within the earth by heat, pressure, and chemically active fluids (Tarbuck and Lutgens 1993:30, 36, 41, 42, 45, 46, 49, 636, 638, 641).

Regardless the type of rocks, all the knappable rocks are composed largely of silicate. The most abundant group of silicate minerals on the continental crust is quartz. As quartz is the only mineral made purely of silicon and oxygen, the term "silica" is applied to quartz (SiO_2). Such stones as most limestone, siltstone, and sandstone that are not composed of silicate are not used by knappers. Limestone is composed chiefly of the mineral calcite (calcium carbonate, $CaCO_3$). Siltstone is an indurated silt that has the texture and composition of shale but lacks fine lamination or fillilty. Sandstone consists of grains of sand size-set carbonate in a matrix or silt or clay. Grains in sandstone are usually cemented by silica, iron oxide, or calcium (Bates and Jackson 1984:295, 446, 469). All these three rocks are too soft or unconsolidated to flake well. In contrast, man-made glass, obsidian, chert, basalt, quartzite, rhyolite, and quartz are preferred by knappers. These rocks are characterized by stone quality suitable for flaking which is mentioned above. Each of these rock types will be discussed in the following section.

Igneous Rock: Obsidian. Obsidian is a common type of natural volcanic glass and is included in igneous rocks. Obsidian is characterized by vitreous, luster of glass, and isotropic properties. Obsidian is generally black or reddish-brown in color due to the presence of metallic ions. Textures vary from homogeneous and glassy to crystalline with a grainy or sugary texture. It consists largely of silicas such as potassium feldspar and quartz as it is chemically related to rhyolite and granite. However, the lack of the formation of distinct crystals through extremely rapid cooling of the molten rock differentiates obsidian from its chemical relatives (Tarbuck and Lutgens 1993:67). Although obsidian appears in large massive flows and as lumps or beds, it is quite difficult to quarry pieces from the original deposits. Prehistoric knappers commonly collected obsidian from secondary deposits like talus slopes and streambeds where it ended up in the form of chunks or nodules (Whittaker 1994:69). Because obsidian responds to the workers' intent and gave flakes, blades, and tools with a keen edge, it was universally preferred by toolmakers for certain tools (Crabtree 1975:108).

Igneous Rock: Basalt. Basalt is composed primarily of pyroxene and calcium-rich feldspar with lesser amount of olivine and amphibole. Basalt contains small, light-colored calcium feldspar phenocrysts or glassy-appearing olivine phenocrysts embedded in a dark groundmass if it has a porphyritic texture. A porphyritic texture refers to an igneous rock texture that is characterized by two distinctively different crystal sizes. The larger crystals embedded in a matrix of smaller crystals are called phenocrysts, whereas the matrix of smaller crystals is termed the groundmass (Tarbuck and Lutgens 1993:69, 640). Basalt is dark gray to black. Basalt forms the floors of modern oceans and has too small crystals for individual minerals to be distinguished with the unaided eye. Because basalt contains a high percentage of ferromagnesian minerals, geologists refer to it as mafic rock (Busch 1993:63). Depending on the rate of magma cooling, the flaking qualities of basalt are variable ranging from fairly homogeneous, to coarsely grainy, to completely unflakable. In general, basalts are quite tough and not a good material for making fine tools (Whittaker 1994:69-70).

Igneous Rock: Rhyolite. Igneous rock rhyolite consists primarily of the light-colored silicates such as quartz and potassium feldspar. It is usually buff to pink or occasionally very light gray in color. Although rhyolite and obsidian have same chemical composition, their property is different. In the case of obsidian, there is no time for mineral crystals to form because lava or magma is cooled abruptly. On the other hand, rhyolite consists of very small crystals, called aphanitic texture because it is formed by magma or lava cooled slowly relative to obsidian. Like basalt the flaking qualities of rhyolite varies. Rhyolite is not a good material for making fine tools.

Sedimentary Rock: Chert. Cryptocrystalline silicates include chert, flint, jasper, chalcedony, agate, and onyx. Many archaeologists usually subdivided the general category of chert into varieties such as flint, jasper, chalcedony, and agate. Some people call red cherts as jasper, and translucent cherts as chalcedony. However, as to the definition of these categories, there is no consensus among geologists. In the present study, chert, jasper, and chalcedony are categorized as chert.

Chert is chemical sedimentary rock that is originated from soluble material produced largely by chemical weathering. Chemical sedimentary rock is formed from such dissolved substances precipitated by either inorganic or organic processes. Thus, a number of very compact and hard rocks composed of microcrystalline silica (SiO_2) are generally called as chert. Cherts consist primarily of silicon and oxygen with any other element as a contaminant or impurity. The impurities occur generally in association with clay, iron oxides, carbonates, or organic matter. Flint is a variety of chert characterized by dark color caused by the organic matter content. Flint and chert have the same chemical composition and structure. Chert is usually found either as irregularly shaped nodules in limestone or as layers of rock. As the silica, the major constituent of chert, is deposited directly from the water, chert nodules do not contain organic matter (Luedtke 1992:7, 17, 46; Tarbuck and Lutgens 1993:138, 145). Most nodules generally have a rough and calcareous surface, called cortex. The finest cherts have smooth and glassy fracture surfaces. Nonetheless they are tougher than glass or obsidian. The quality ranges from fine material characterized by cryptocrystalline but not glassy, tough grainy stuff, to extremely coarse material with visible crystals. Most of the flints and cherts in prehistory came from secondary deposits. Chert nodules are often eroded out of the parent bedrock and incorporated in sedimentary deposits such as river gravels or exposed on the surface as they are usually more durable than the surrounding material (Whittaker 1994:70-71).

Chalcedony. Although chalcedony is made of quartz, its crystals grow as radiation fibers in bundles and its structure is more porous than chert which forms grains. Chalcedony crystals grow in a viscous solution that removes pure silica from the liquid and rejects any impurities. Consequently, most chalcedony is translucent and pale. Agate is chalcedony that has different colors arranged in strips or layers (Luedtke 1992:24-25). Thus, chalcedony is a chert variety characterized by the its purity of its silicates (SiO$_2$).

Jasper. Jasper is made of cryptocrystalline silicates like other chert variety such as flint, and chalcedony. Archaeologists call red-, reddish-brown-, and mustard-colored cherts jasper; geologists generally limit the term jasper to cherts that contain high iron oxide impurities.

Sedimentary Rock: Sandstone. Sandstone is the name of rocks that consist primarily of sand-sized grains. Sandstone is detrital sedimentary rock which is formed by accumulation of materials that are originate and are transported as solid particles derived from both mechanical and chemical weathering. Typical sand is composed largely of quartz with lesser amount of feldspar, mica, garnet, magnetite, and other minerals (Busch 1993:79, 83, 84). Sandstone is generally used for grinding and polishing. However, very compact and homogeneous variety can be used as various artifacts by percussion technique (Crabtree 1972:89, 90).

Metamorphic Rock: Quartzite. Quartzite, formed from quartz sandstone, is a very hard metamorphic rock. Recrystallization is so complete that quartzite splits across the original quartz grains when broken. Quartzite is usually white, but iron oxide may produce reddish or pinkish stains. Dark mineral grains cause a gray color. Most metamorphic rocks are characterized by foliation, which refers to a linear arrangement of textural features. Because of the high content of quartz, quartzite lacks a strong metamorphic foliation (Tarbuck and Lutgens 1993:170). Most of the quartzite used by prehistoric knappers was found in the form of cobbles in riverbeds. Most quartzite is tough stuff and usually produces crude looking tools. However, quartzite pebbles are very good hammer stones for cherts and other hard materials (Whittaker 1994:72).

Mineral: Quartz. Quartz is the only common silicate mineral composed entirely of silicon and oxygen. Quartz is an important rock-forming mineral that occurs either as transparent hexagonal crystals or as macrocrystalline or microcrystalline grains in rocks (Luedtke 1992:153). Because quartz is bonded by a three-dimensional framework developed through the complete sharing of oxygen by adjacent silicon atoms, quartz is very hard, resistant to weathering, and does not have cleavage. When broken quartz generally exhibits conchoidal fracture (Tarbuck and Lutgens 1993:49-50). In a pure form, quartz is colorless. However, like most other clear minerals, quartz is often colored by inclusion of various ions. The most common varieties of quartz are milky (contains microscopic fluid inclusion such as water), rose (contains titanium), violet (contain ferric iron). In addition quartz can be gray, blue, green, red, yellow, brown, and more

(Busch 1993:27-28). Like quartzite, quartz pebbles are very good hammer stones for cherts and quartzite.

The Mechanical Process of Fracture

In archaeology, the study of stone tools is important because it provides information about the cultural development of prehistoric man. The reconstruction of cultural development is generally based upon the trace of manufacturing techniques and wears on tools because they provide some insight into ancient technology and behavior. In this course, understanding the fracture mechanics are important because it provides us clue to identify manufacturing techniques and elucidate the nature of use-fracturing (Lawn and Marshall 1979:63; Cotterell and Kamminga 1979:97). As mentioned in the previous chapter, the principle mechanics of flake formation in stone tool making and use wear are the same and different only in the scale. Therefore, fracture mechanics is the foundation of lithic technology and provide insight into human behavior.

Flake Type

The great variety in the morphology of human-made flakes; flakes can be grouped into three general types: conchoidal, bending and compression flakes. The conchoidal flake is named by the appearance of half bivalve shell inside surface of some flakes given by distinctive bulb of force and concentric undulation on the fracture surface. A conchoidal flake is only formed by a comparatively hard indenter, such as hammer stone. The bending flake is characterized by the unintentional breakage related to bending. A compression flake, initiated by microscopic wedging, is produced during bipolar flaking (Cotterell and Kamminga 1987:684-685). In general, the formation of flake follows three sequential phases: initiation, propagation, and termination.

Initiation Phase

In fracture mechanics, initiation refers to how and where cracks start. All flakes can be initiated on the basis of one of three basic modes: Hertzian, wedging, and bending initiations. In conchoidal flaking, a formation of flake is initiated by the formation of a partial Hertzian cone crack around the contact zone between the flaking tool and the initiation surface on the core. A wedging initiation occurs directly under the applied force. This initiation is largely associated with bipolar flaking. A bending initiation appears relatively far from the area of contact between the core or material worked and a flaking tool (Tsirk 1979:84-85).

Hertzian Initiation. A cone crack develops when a force is applied by a hard radial indenter to a brittle elastic solid, and the applied force exceeds some critical value. This cone crack is called as a Hertzian cone fracture naming after the German physicist Heinrich Hertz (1881-1882)

21

who first studied the crack pattern produced when two curved bodies were brought into mutual contact. He described the way in which the elastic contact circle expanded with applied force until indenter caused the sudden development of a cone-shaped crack, the so-called Hertzian cone crack in the brittle solid (Speth 1972:34). A classic Hertzian cone fracture is developed when a bodies of hard smaller radius, which generates more intense contact pressures at a loading, indents perpendicularly into the flat surface of an isotropic brittle solid. While the contact stress in the surface is compressive under the contact area between the indenter and the solid, near the edge of the contact zone where the stress is highest becomes tensile. On further loading of indenter, the contact circle expands and the symmetrical surface ring crack is driven downward until it reaches a critical size. At this moment the ring crack develops as a visible cone. As the load increases, the crack continues in stable extension until the contact circle encompasses the surface trace of the cone, whereupon the primary crack closes up within the engulfing compressive zone (Lawn and Marshall 1979:67-68). In actual lithic practice, the initiation of a conchoidal flake does not follow exactly those required for a classic Hertzian cone. Initiation of conchoidal flakes usually occurs near the side face of the core and the initiating force generally has an outward component. As a result, the tensile radial stress in the contact zone furthest from the side face is enhanced, while those near the side are changed to compression. These compressive radial stresses near the side face thus lead to formation of partial Hertzian cone. To solve this problem, the necessary high pressure is required. In other words, conchoidal flakes can be formed when the indenter is hard. Unlike bone or antler pressure-flaking implements, wood is usually too soft to produce a conchoidal flake by pressure but sometimes it can produce conchoidal flakes by percussion. As mentioned earlier, if a striking platform is located just above the ridge on a core, a soft hammer produces flakes with strong bulb of percussion. The partial Hertzian cone determines the shape of the bulb of force, which is a characteristic of the conchoidal flake (Figure 11). A hard hammer often produces a number of concentric partial Hertzian cone cracks as initiation face of the cores. One of these cracks results in the formation of the flake (Cotterell and Kamminga 1987:686-687).

Wedging Initiation. When longitudinal stress by a very hard sharp indenter strikes a surface of the brittle solid perpendicularly, the wave travels the core perpendicularly and the surface of the core deforms plastically (Speth 1972:38). At this moment incipient flaw of a solid in the subsurface zone of maximum stress concentration is generated by the plasticity process. If the load become critical, couples of deformation-induced flaws grow into small subsurface median cracks on symmetry planes that contain the load axis and the major impression diagonals. As the load increases, the contact expands and drives the median cracks steadily downward into the brittle solid and

upward to intersection at the free surface. When the load is withdrawn, the walls of the median fractures try to join together. However, fractures of debris, acting as wedging mechanism, prevents complete join. Even after the indenter leaves the surface completely, the lateral cracks continue to spread or grow (Lawn and Marshall 1979:71-76). This wedging initiation is predominant in bipolar flaking.

Bending Initiation. Unlike the Hertzian initiation, bending initiation occurs relatively far from the area of contact between the brittle solid and an indenter. Thus, if the fracture is initiated by the bending, there are no cone features. In Hertzian initiation, a decrease in the size of the contact area results in a significant increase in the tensile stresses critical for fracture initiation near the contact area, while a decrease in the size of contact area does not affect the tensile stresses critical for bending initiation because initiation appears relatively far from the contact area. Thus, when the bending initiation occurs, the maximum stress on the initiation face appears away from the point of application of force. Particularly, the highest bending stresses occur on edges with small angles (Tsirk 1979:82-90).

In general, bending initiation is caused by a soft indenter such as bone, antler, wood or soft stone including limestone. Particularly, biface manufacture commonly produces bending flakes. However, if the angle between the surface of a core and the striking platform is small, hard hammer percussion can also produce bending flakes. Flakes initiated by bending show such characteristics as a concave scar on the initiation face, no bulb of percussion, and a conchoidal flake with pronounced lip. As bending initiation does not produce secondary detachment, the resultant flake can be refitted to its scar without any appreciable gap. Flake initiation in preforming and in rejuvenating worn edges by soft-hammer percussion and pressure flaking with wood, antler, or bone are usually entail bending (Cotterell and Kamminga 1987:690-691). Use wear, flaking damage in the form of bending initiated scars on a small scale is commonly produced by adzing wood, sawing bone, and cutting activities (Lawrence 1979:118-119; Odell 1981:199).

Propagation Phase

Propagation refers to the paths that the fracture follows and the parameters that determine the extent of their growth (Lawn and Marshall 1979:66). After the fracture has left the initiation stage, it has entered the propagation stage. The crack or fracture in any solid propagates either steadily under mechanical work done by external forces or unstably under the release of elastic strain energy. In general, the stable direction for fracture propagation depends on force angle in flaking and load rate. Therefore, subsequent path of fracture becomes uncertain. However, directional

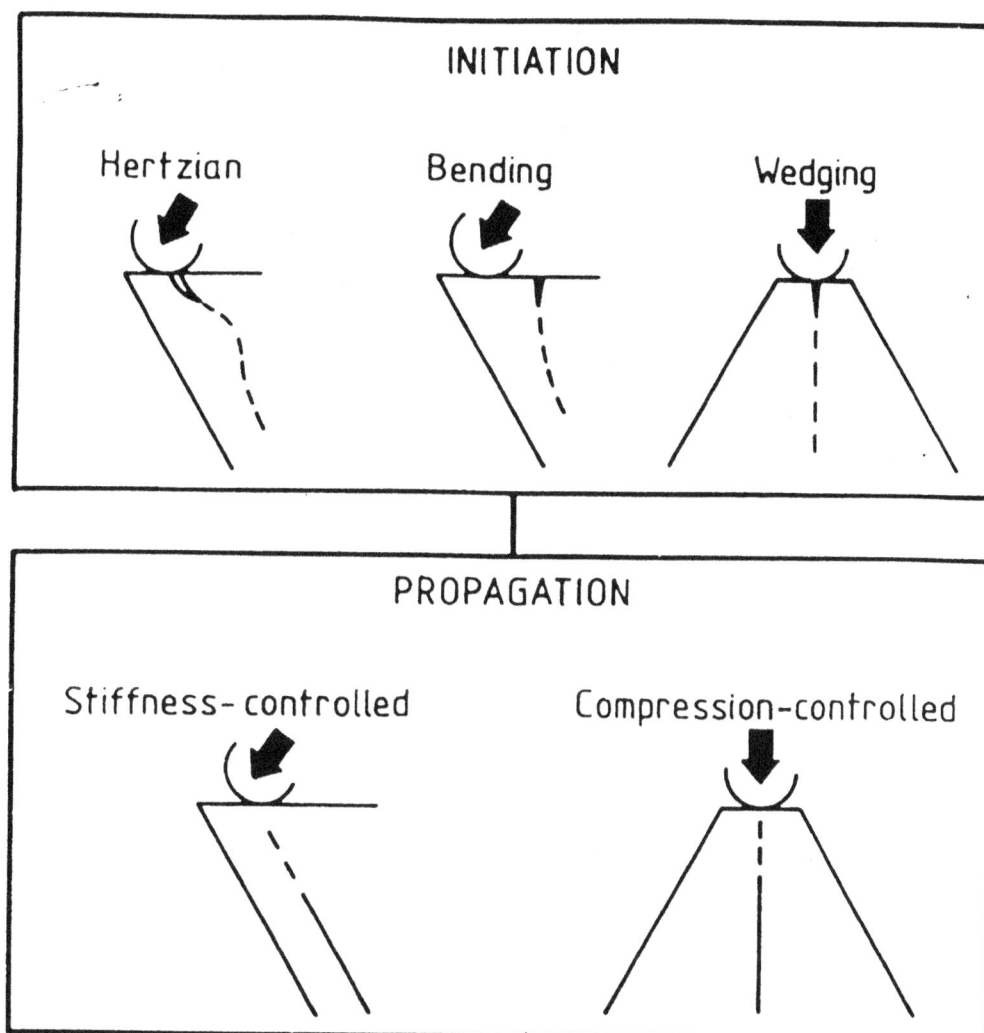

Figure 11. Initiation and Propagation.

stability of fracture can be achieved if sufficiently large compressive stresses exist that are oriented in the direction of the fracture. There are two types of crack propagation in flaking. They are the fractures propagated under a combination of bending and compressive forces and the fractures propagated under secondary tensile stresses produced by compressive stress fields far away from the edge of a core. In the former case, the crack path has been controlled by the stiffness of the flake and the force angle is the most significant parameter. In the latter case, the crack path has been controlled by compression and the fracture propagated under tensile stresses is closely related to bipolar technique (Cotterell and Kamminga 1987:692).

Stiffness-Controlled Propagation. External stress applied to a solid material consists of minimal stress and maximum stress. The direction of the minimum principal stress with the maximum principal stresses acting across the fracture surface is the stable direction for fracture propagation. Conchoidal and other flakes formed along the side of a core are usually long relative to their thickness because the

fracture propagation is parallel to the side face of a core (Cotterell and Kamminga 1979:103) although the flake length is also affected by the angle of exterior platform and platform thickness (Dibble and Whittaker 1981; Speth 1972, 1974). When bending initiation occurs at the edge of the contact zone during the formation of a flake, the principal stress trajectories is radically altered by the advancing fracture. At this moment, only the maximum principal stress trajectory passes through the tip of the fracture and the direction of the fracture becomes unstable. Consequently, subsequent path of fracture becomes uncertain. However, directional stability of fracture can be achieved in this case if sufficiently large compressive stresses exist that are oriented in the direction of the fracture. When the angle of the force, the angle between the flaking force and the side face of a core, can be increased or the effective bending can be reduced, such compressive stresses can be achieved (Cotterell and Kamminga 1979:103). Therefore, there is a close correspondence between the flaking angle and the angle of the force determined by flake stiffness in the removal of long thin flake. After the propagation phase, the angle of the force which is important parameter of the propagation phase does not affect the variables of flake termination, while it little affects interior platform angle which is the flake's initiation angle (Dibble and Whittaker 1981:295).

Compression-Controlled Propagation. Flakes formed away from a side face of a core propagate under compression in the direction of the stress. Initiation of flake is related to wedging which contributes to the crack's growth (Cotterell and Kamminga 1987:698). When the force is directed into the body of the core, a primary compressive stress is produced. At this moment, the principal stress trajectories near the tip of an advancing fracture are distorted. However, the prolongation of the flake still follows the direction of the minimum principal stress that is normal to the surface. Therefore, the fracture under stable direction tends to produce long flakes (Cotterell and Kamminga 1979:103).

Termination Phase

After the propagation phase, flake fracture results in one of four basic flake endings or terminations: feather, hinge, step, and overshot termination (Figure 12).

Feather Termination. A feather termination is produce when the fracture forming the flake has been propagated parallel to a side face of a core and then turns slightly to meet the side face at a very acute angle. The feather termination is the natural termination for the stiffness-controlled propagation phase of a flake. The significant parameter in the formation of feather termination is low exterior platform angle. In use wear analysis fine terminations on feather scars are not important because most of use wear results in either hinge or step fractures (Dibble and Whittaker 1981:287; Cotterell and Kamminga 1987:699). In general, sharp feather terminations are desirable because they produce a sharp edge on the flake and a smooth surface on the core. Correct flaking largely results in feather termination (Whittaker 1994:106).

Step Termination. Step termination results from an abrupt change in the direction of the crack in a right angle break at the point of truncation. The major cause of this termination is a dissipation of stress resulting from insufficient energy available to complete the fracture or from the collapse of the flake due to a material flaw that effectively blunts the fracture (Crabtree 1972:93; Dibble and Whittaker 1981:287). Step termination is a common feature on exhausted cores of poor quality lithic materials such as quartzite. When a hammer stone strikes the core, there exists applying force in two directions. One is a downward force that compresses the material and leads the fracture to the core in the direction of the flake's length and the other is an outward force that pulls the flake away form the core when the crack between core and flake lengthens and widens. At this moment, if the outward force exceeds the downward force, the flake is pulled away

from the core too rapidly and snaps when it is bend beyond its endurance (Sollberger 1986:101-105).

Hinge Termination. A hinge termination is a crack at the distal end of a flake that prevents detachment of the flake at its intended terminal point. A hinge termination occurs at right angles to the longitudinal axis. The break end is rounded or blunt in cross section (Crabtree 1972:68). Like step termination, a hinge termination is formed when more outward pressure exists than is expected or required to produce a feather termination. Here the reduction in inward compressive stress is accomplished by a too light indenter in weight or by too little force. The major reason for the reduction in compression is deflection and rotation of the flake by the indenter. When the loss on compression occurs by the alteration of the angle of the indenter, the compressive fracture begins to convert to a fracture in bending. As a result, the width and/or thickness of the flake is increased relative to the established fracture trajectory. When this situation occurs, proximal part of the flake is more widely separated form the core and the indenter produces a lever that causes secondary compression at the position of the distal end. This secondary compression causes tension directed toward the proximal end. As fracture extends, the distal part forms a material resistance to fracture propagation. At this point the primary compressive load of the advancing fracture front become diminished to a level that precludes it from exiting directly outward as its lateral force is insufficient to penetrate the secondary compression barrier (Figure 12). After the fracture passes the bending fracture of the distal tip, the load is compressed at the bending. Compression at the bending causes the fracture to extend upward. Contact of the distal end of the flake at bending finally dissipates all tension upward as well, leaving only forces of outward bending at termination. When the fracture front is very close to the outside of the core, the final termination of the flake ends with fracture-in-bending (Sollberger 1994:17-20).

Plunging or Overshoot Termination. When the exterior platform or edge angle of the side face of a core is nearly 90 °, the flake can fail to develop completely or is said to overshoot. In most cases, such flakes plunge towards the interior of the core along the path that the fracture exited the core. That is, from the side opposite the flaking surface and not the side opposite the striking platform (Dibble and Whittaker 1981:287). Although overshoot termination occurs in some forms of use wear, such as scraping wood, it is most common in blade flaking. When a blade is removed from a blade core, plunging termination is produced if the position of the tip of the pressure-flaking implement is located too far from a side face of a core. This is because the crack propagates toward the tip of the core. However, thin blades generally end with a feather termination as blade cores are shaped with a gentle curvature at their base (Cotterell and Kamminga 1987:701).

Feather termination

Hinge termination

Step termination

Overshoot termination

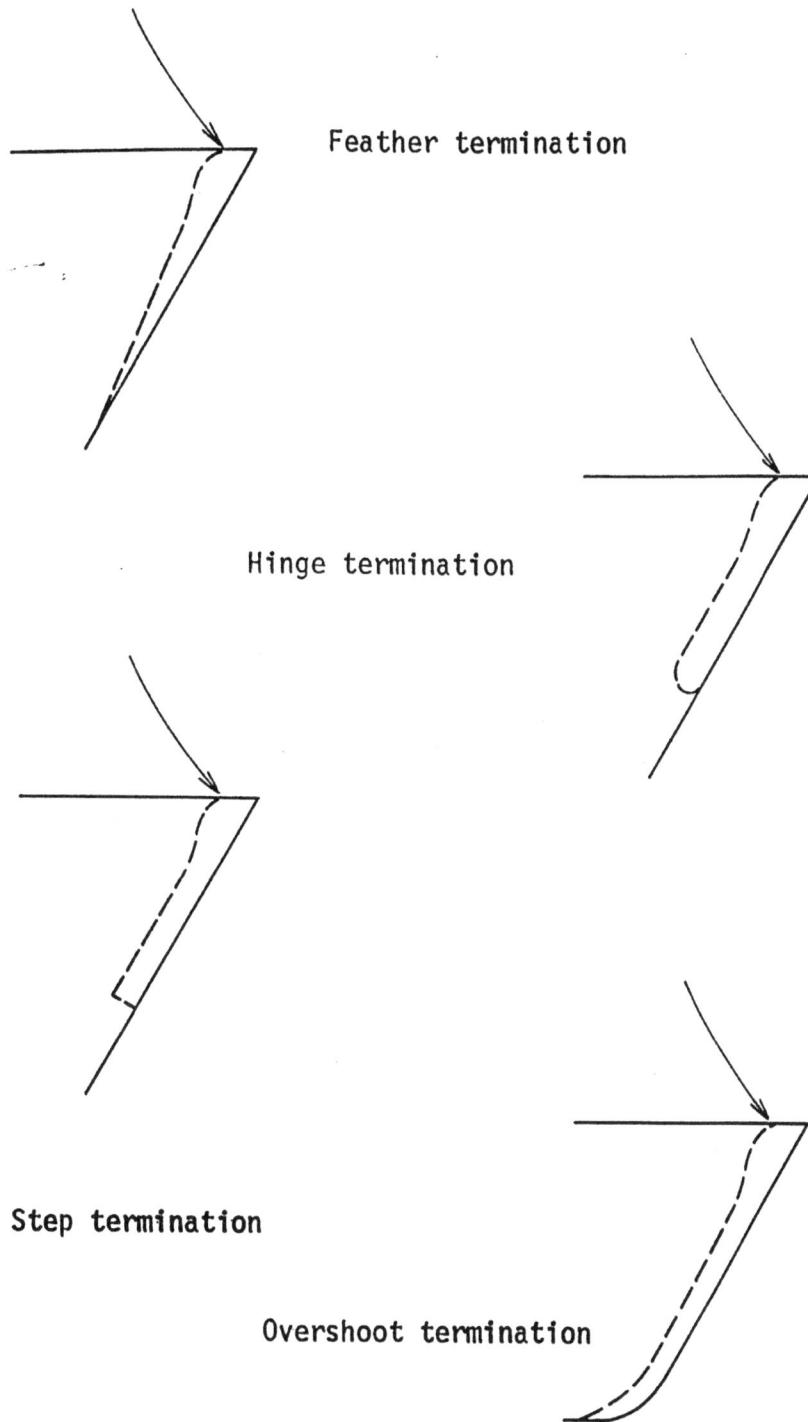

Figure 12. Types of Termination.

Reduction Techniques

Analysis of the lithic assemblage in this present study reveals that several modes of lithic techniques were used. Based upon the attribute analysis, the use of hard hammer and soft percussions, bipolar technique, pressure technique can be demonstrated. In addition, Heat treatment and battering technique are present. In this section, I describe each technique that was applied to produce flakes in the

study assemblage. However, the number of flakes in each technique was not counted. As mentioned earlier, although, in general, flakes produced by a certain technique carry special morphology on it for instance, hard hammer percussion produces pronounced bulb of percussion, soft hammer percussion diffused bulb, bipolar technique severed bulb, and so on these are general associations, not invariant ones. Some times artifacts are produced by a combination of multiple individual techniques. In other instances, artifacts do not exhibit diagnostic features of the technique. Finally, different techniques can produce artifacts that are identical morphologically. However, these morphological characteristics, in general, provide us information of the presence of a certain technique at the site.

Percussion Techniques

Percussion is the oldest method of working stone (Semenov 1964:39). In percussion techniques, the implement or fabricator must be large and heavy enough to induce sufficient force in order to exceed the elastic limits of the stone to initiate fracture or crack. In general, hammer stones of different hardness, texture and size, billets or rods of wood, antler, bone, ivory and horn are used. The type of hammer stone material and the technique depend on the quality of raw material being worked and the stage of manufacture (Crabtree 1972:8-9).

Direct Freehand (Hard-hammer) Percussion. Hard-hammer percussion is an extremely simple technique, which is the most widely used one at the earliest archaeological sites. This technique consists of holding an objective rock by one and striking vertically to the margin of the rock with a hammer stone or billet by the other hand to remove flakes. The major feature of artifacts made by this technique carries pronounced bulb of percussion and sharp edges. After several flakes are removed from a core, a core can be used as a chopping tool. If the core is turned after each flake is removed and the previous flake scar is served as a striking platform for the next flake removal, a bifacial tool such as a simple pointed handaxe can be produced. The handaxe was used for a multi-purpose implement, including cutting, chopping, boring, and digging (Crabtree 1972:11). In this technique, the suitable hammer stone is usually a fairly smooth cobble that has a rounded surface on at least one end or side. The preferred hammers are generally oblong shaped. The size of the hammer stone depends on the size of a core. If the core is a pound-size, the hammer stone is about the same size. If a small cobble, it is somewhat larger than the core. If the core is very large, the hammer stone tends to be smaller than the core in order to be held in the hammering hand (Schick and Toth 1993:118).

The significant parameters to obtain more desirable flakes include a suitable piece of a core material, platform angle, flake width, flake length, flake thickness, curvature, the face of a core, the angle of blow or force, and termination. Relations between these parameters are discussed in the previous chapter (Dibble and Whittaker 1981; Schick and Toth 1993:119,121; Speth 1972, 1974; Whittaker 1994: 93-116).

Bipolar Percussion. Bipolar percussion is a process that involves a core being struck straight downward from above, perpendicular to both the core top and the anvil. Although theoretically the bipolar percussion produces a sheared cone and a flat ventral surface on the flakes, this technique can produce a variety of flake types. The wide variety of shapes depends on whether (1) a core is a block, a nodule, a cobble, or a pebble, and (2) the core is worked from the outside in or inside out, and (3) the flake splits perpendicular, parallel, or at odd angles to the plane of the core face (Callahan 1994:15, 16, 31, 34). Bipolar technique is largely used for rock types such as quartz and quartzite that are very hard to produce flakes in the hand because of their unpredictable fracture pattern.

Soft-hammer Percussion. Soft-hammer percussion, also called billet or baton technique, appears later than the hard-hammer percussion in the Stone Age. The hammer in this technique is much softer and more resilient than the stone core. Major features of flakes or tools produced by this technique involve diffused bulb of percussion, the presence of lipping of the interior of the platform, very thin, and usually arched in cross-section (Crabtree 1970:150). The fracture theory of the bending is closely related to the soft-hammer percussion. The core also yields shallower and less well-marked scars on it. Usually antler, bone, ivory, and wood billet are used in this technique. Soft-hammer percussion is often applied to thinning, flattening, and shaping bifaces because it is the easiest way of removing large, relatively flat and thin flakes without pronounced bulb of percussion. The flakes produced by the soft-hammer technique also tended to expand in width from the platform (Whittaker 1994:185-199).

Indirect Percussion (Punch Technique). This technique useful for making and finishing tools requires the use of punch or blunt rod-like implement of stone, bone, antler, horn, ivory or hard wood like dog wood. The quality of the material being worked and the technique determine the punch type. The most distinctive advantage of this technique is the maintenance of constant angle during the percussion and production of more straight and uniform flakes with small platforms because the tip of the punch on the platform is located accurately. However, holding the preformed core is very difficult. As a result, two people are required using this technique (Crabtree 1972:12-13).

Pressure Technique

Pressure technique is performed holding the objective piece in one hand and pressing the objective with some organic materials such as bone, ivory, hardwood, horn, shell, and so on. The pressure technique varies depending on the worker's design of the artifact, combinations of inward and outward pressure, methods of holding position, and characteristics of raw material. In right-hand person case, the objective piece is held at a right angle on a pad in the cupped palm of the left hand resting the left wrist inside of the left thigh, giving additional support. The tip of the pressure tip is seated at a right angle to the long axis of the objective piece on the leading edge. By removing the flakes from the piece, the edge has been centered. In this process, a platform is established by removing small flakes or by abrading to prevent crushing. Flakes are detached from the surface, which is faced toward thumb. In general, flakes produced by the pressure technique are relatively small and thin compared to percussion. However, there are exceptions as well. Two methods of using the cone principle can be applied to pressure. As pressure force is applied in only a downward direction, the flake is removed tangentially to the direction of the applied force like the direct percussion. These flakes are characterized by small, feather termination, and salient bulbs of pressure. However, if an applied pressure travels along the ridge or convexities, long, and narrow flakes can be produced. Pressure technique is useful for shaping, tool making, sharpening, modification, and preparation (Crabtree 1972:15-16).

Heat-treatment (Shurmai [GnJm 1]: N' 131, Kakwa Lelash [GnJm 2]: N' 107)

As briefly discussed in the previous chapter, heat-treatment can improve the quality of some cryptocrystalline silicates. On a general level, the changes that occur in thermally altered materials are diverse. Some cherts show extreme changes, some are unaffected, and some are completely destroyed. Materials like obsidian, basalt, rhyolite, quartzite cannot be heat-treated at all (Whittaker 1994:94). The most frequently used material by the heat-treatment is chert.

When chert is heated at temperatures around 250 ° C, many cherts show a pinkish coloration that is not present in unheated material. Dehydration, presence of material or organic impurities, and oxidizing vs. reducing heat environment are suggested for the reasons for these color changes. When chert is fractured after it was heated at temperatures about 350°C, fracture surfaces are characterized by a greater degree of luster than the old ones. The degree of luster increase is positively correlated with the duration of peak temperature. However, how much luster is increased varies among different silica materials. Successful heat-treatment of raw material requires a high enough temperature and slow heating and cooling. If the heat increases too fast, or if the heated flakes are exposed suddenly to cold air, crack, shatter, or potlid which is little round flake that pop off the surface and leave an irregular pitted scar occur (Purdy and Brooks 1971:322-325; Rick and Chappel 1983:69).

Heat-treatment also improved working quality. Heat-treatment increases fracture strength. Compared to raw chert, thermally altered chert produced flakes with less hinge fracture and generally longer flakes. In addition, heat-treated flakes are significantly thinner and lighter in weight (Mandeville and Flenniken 1974:147-148; Patterson 1979:257-258). The major disadvantage of heat-treatment is that thermally altered materials are weaker and more likely to fracture (Whittaker 1994:73).

Battering (Shurmai [GnJm 1]: N' 3, Kakwa Lelash [GnJm 2]: N' 5)

Battering results from use rather than intentional retouch. If cryptocrystalline materials are used as battering numerous precursor or fabricator marks or impact points indicating fractured cones of percussion appear on the material. These materials used as continuous battering results in crushing the outer portion and the end product is generally characterized by a rounded and roughened surface. Although the same result appears on noncryptocrystalline stones, the cones of percussion do not occur on these materials. Therefore, noncryptocrystalline materials are preferred for battering tools (Shafer 1978:71-72).

Chapter IV
Comparative Analysis of Lithic Assemblages from Shurmai (GnJm 1) and Kakwa Lelash (GnJm 2) Rockshelters, Kenya

In this chapter I discuss raw material procurement and technological strategies and reconstruct the core reduction sequence evident at the sites of Shurmai and Kakwa Lelash rockshelters. I also discuss function of the tools that were recovered at these two sites. The functional analysis consists of measuring angle of modified edge and use-wear analysis. Finally, the style of these tools is discussed.

History of Present Research

Material excavated from Shurmai (GnJm 1) and Kakwa Lelash (GnJm 2) rockshelters during the summers of 1992-1995 were carefully sorted out as lithics, pottery, bone and other organic material in the field. As required by Kenyan law, excavations of these sites were conducted in affiliation with the National Museums of Kenya. According to the National Museums of Kenya policy for material excavated in Kenya:

> All specimens collected have to be deposited at the Museum. Accessioning and storage will be undertaken by the Curatorial staff of the Division, but all specimens have to be labeled and catalogued, and counted according to specific categories for computerization as indicated by the Curatorial Staff (National Museums of Kenya 1993).

Following this regulation, all excavated material from Shurmai and Kakwa Lelash rockshelters was transported to the National Museums of Kenya in Nairobi where it was washed, labeled, and curated.

I spent approximately eight weeks from June 24 to August 21, 1996 conducting an analysis of the lithic assemblage excavated from these two sites. This work was undertaken in the laboratory of Division of Archaeology of the National Museums of Kenya in Nairobi because of the following regulations issued by the Museum.

> All specimens are and shall remain the property of the Kenya Government and are controlled by the Antiquities and Monuments Act, 1983. Specimens are not allowed to be temporarily exported for study, expect under special circumstances. Researchers are expected to conduct their studies on materials in Kenya and should plan their research schedules accordingly. Dating samples and materials which require analyses, for which facilities are not available in Kenya, will be allowed to be exported. Applications for export of archaeological research materials should be submitted after consultation with the Head of the Division to the Director/Chief Executive, National Museums of Kenya (National Museums of Kenya 1993).

However, because of time constraints and the absence of a scale for weighting lithic material, I could not finish my analysis in Kenya. Dr. Karega-Munene, Head of the Division of Archaeology of the National Museums of Kenya, kindly arranged for me to ship the remainder of this lithic assemblage from Kenya to Texas A&M University in order to finish the work.

The same methods of lithic analysis were used at both the National Museums of Kenya and Texas A&M University. The lithic assemblage excavated from Shurmai Rockshelter (GnJm 1) consists of 4,782 stone artifacts, while the Kakwa Lelash Rockshelter (GnJm 2) produced 7,862 stone artifacts. The assemblage excavated from these two rockshelters thus totals 12,644 lithic objects categorized as flakes, tools, cores, or debris. Definitions of each category are provided in the Chapter VI. As part of my analysis I categorized all stone artifacts except for debris by means of a four-digit hierarchical numbering system, and coded their attributes using a numeric system. For reasons described previously, the only lithic materials not coded in this manner were flake fragments.

The primary purpose of this system of data recording was to create a computerized file containing all the collected information for each artifact catalog number assigned. This computerized data file was created for easy analysis of this information using existing statistical and graphical packages. This coding system allowed me to undertake both multivariate and univariate analysis. Fox Pro database program was used for the statistical and graphical analyses of all collected data. The statistical analyses utilize the Simstat program for chi-square and t-test. The data were processed at the Department of Anthropology, Texas A&M University.

In conducting this analysis, I had three primary aims: (1) to describe the attributes of debitage, tools, and cores from the two sites, (2) to measure and compare the technological change through time and style and function of tools at the two sites, (3) to monitor and compare changes in raw material usage at each through time, and (4) to correlate my research with environmental change. Further, I wanted to use these three classes of data to help me understand human behavior in the Middle Stone Age in Africa in general.

Raw Material Analysis

Lithic Raw Material Procurement Strategies

The acquisition of good raw material for flaking in prehistory would have been a critical problem. No doubt some regions had rich stone sources while others did not.

In the former case, people did not need to procure stones in places far away from their settlements, in the latter, people had to acquire their materials from a distance. The study of raw material used for the manufacture of stone artifacts, therefore, provides us information about prehistoric peoples' exploitation of their environment.

As noted, the Shurmai (GnJm 1) and Kakwa Lelash (GnJm 2) rockshelters are located on the top of large Pre-Cambrian metamorphic granite gneiss erosional remnants or inselbergs. Because of their geological settings, the sites contain rich volcanic rocks, such as basalt, granite, gneiss, and schist. In addition, the sites are in a rich quartz source area.

Although these two archaeological sites have abundant stone sources, many of rocks that are found nearby are not good raw material for flaking or making fine tools. As a result, the inhabitants of the sites were faced with a choice: procure raw material far away from their site or utilize the inferior local raw material instead. In order to understand the raw material procurement strategies, the raw materials produced by two dry rockshelters were categorized as local, non-local, and exotic materials. Following Brown (1991), materials which are found within 10 kilometers from the site are categorized as *local*, those found between 10 and 50 kilometers from the site are *non-local*, and those found more than 50 kilometers from the site are *exotic* materials.

Raw Materials Utilized at the Shurmai Rockshelter (GnJm 1). Lithic debitage assemblage produced by Shurmai Rockshelter consists of 75.4 percent local material. The dominant material of this assemblage is fine-grained basalt (68.2 percent) which is found today throughout this portion of the Kipsing River drainage surrounding the site. Most of these basalts are severely weathered and make mediocre to poor knapping material. Most of the cores recovered there were of basalt. Together, local basalt and quartz comprise 85 percent of cores in the assemblage.

Non-local materials constitute a significant component of unmodified flake assemblage at the Shurmai Rockshelter. This material constitutes 21.2 percent of the assemblage. The dominant non-local material is chert. Most of chert found in the site is good knapping material. Exotic materials comprise only 3.5 percent of assemblage in the site. The only exotic material found in the site is obsidian.

Five raw material types were used for manufacturing stone tools. They are basalt, chert, obsidian, quartz, and sandstone. Like other assemblages, local materials are the predominant materials utilized. These materials constitute 63.3 percent of tool assemblage. However, non-local material is also significant component of the tool assemblage. These materials comprise 24.0 percent of the tool assemblage, while exotic material 12.7 percent. Thus, both non-local and exotic materials form important components of the tool assemblage. These analyses suggest that the stone knapping at the Shurmai (GnJm 1)

Rockshelter was heavily dependent upon a procurement strategy that emphasized local material (Table 2). However, it is assumed that the raw material procurement strategies of the Shurmai Rockshelter would change between the MSA and the LSA occupations of the site.

Of the four units that comprise the stratigraphy of the *sondage*, Unit 2 yielded the majority of local materials in the Shurmai Rockshelter assemblage. Unit 2 produced a thermoluminescence age of 45,211 ± 5356 yr B.P., Unit 3 a radiocarbon age of 20,000 ± 80 yr B.P. (Beta 85593), and Unit 4 yielded a radiocarbon age of 1100 ± 90 yr B.P. The lithics from Unit 2 consists of 92.4 percent of local raw materials, 6.7 percent of non-local and 0.8 percent of exotics (Table 3). The dominant material is basalt. It constitutes 85.9 percent of the Unit 2 lithic assemblage. It would appear that non-local and exotic materials were not exploited very much by the occupants of the site during the formation of Unit 2.

In Unit 3, local materials comprise 69.3 percent of the lithic assemblage. The dominant local material is still basalt, which contributes 63.8 percent of all local materials. Non-local materials comprise 25.1 percent and exotic materials contribute 5.5 percent of the lithics from Unit 3.

Compared with Unit 2, the percentage of both non-local and exotic material increased. However, local materials still constitute the major component of the assemblage.

Local materials comprise 18.0 percent of the unmodified lithic flake assemblage of the Unit 4. Unlike other units, local material quartz (10.2 percent) became a major component of the local raw materials procured by the occupants of the site rather than basalt (7.9 percent). Non-local materials contribute 70.3 percent to the unmodified flake assemblage of the unit. The dominant non-local material in the unit is chert. Exotic material comprises 11.3 percent of the Unit 4 assemblage. Compared with the previous units, Unit 4 exhibits increased utilization of both non-local and exotic materials.

In the case of lithic cores, Unit 2 contributes 84.0 percent, Unit 3 only 4.0 percent, and Unit 4 12.0 percent of total core assemblage of the Shurmai Rockshelter. Thus, most of cores recovered from the site came from Unit 2. In the Unit 2, 90.4 percent of cores were made of local materials. The dominant material is basalt. No cores of exotic material are found in this assemblage. In the case of the Unit 3, 75.0 percent of cores were made of local material and 25.0 percent of exotic material. Cores made of non-local materials were not found. Although Unit 3 produced exotic core materials, of four units, the Unit 3 contributes only 4.0 percent of the total core assemblage of the Shurmai Rockshelter. Therefore, the presence of exotic core material in this unit is not significant. Unit 4 core assemblage consists of 50.0 percent of local, 25.0 percent of non-local, and 25.0 percent of exotic materials (Table

4). That is, most cores recovered from the site are all made of local materials. Although there is a slight increase in the utilization of non-local and exotic materials as cores through time, this change is not significant. However, this result does not mean that both non-local and exotic material were not significantly utilized. Based on the debitage assemblage, acquisition of non-local and exotic material is significantly increased through time although the total number of the material discovered from Unit 3 and 4 is lower than that of Unit 4. It thus appears that

exotic materials were transported to the site not as raw nodules or spalls but as preformed blanks. The analysis of the tools from the Shurmai Rockshelter results in a very similar pattern to that of the debitage and core assemblages (Table 5). Through time the utilization of local materials in tool manufacture decreased, while the use of non-local and exotic materials increased. However, in terms of over all frequency, local materials such as basalt and quartz remain dominant in all three lithic assemblages from the site.

Table 2. Frequency and Percentage of Lithic Assemblage of Shurmai (GnJm 1) Rockshelter

Material Type	Unmodified Flakes		Cores		Tools	
	Frequency	**Percentage**	**Frequency**	**Percentage**	**Frequency**	**Percentage**
Local material	1105	75.4	85	85.0	50	63.3
Non local material	311	21.2	11	11.0	19	24.0
Exotic material	51	3.5	4	4.0	10	12.7
Total	**1467**	**100.1**	**100**	**100**	**79**	**100**

Table 3. Changes in Frequency and Percentage of Debitage Assemblage of Shurmai Rockshelter (GnJm 1) through Time

Unit	**Local material**		**Non local material**		**Exotic material**		**Total percentage**
	Frequency	**Percentage**	**Frequency**	**Percentage**	**Frequency**	**Percentage**	
4	49	18.0	187	70.3	30	11.3	**99.6**
3	163	69.3	59	25.1	13	5.5	**99.9**
2	893	92.4	65	6.7	8	0.8	**99.9**
1	0	0	0	0	0	0	**0**
Total frequency	**1105**		**311**		**51**		

Table 4. Changes in Frequency and Percentage of Core Assemblage of Shurmai (GnJm 1) through Time

Unit	Local material		Non local material		Exotic material		Total Percentage
	Frequency	**Percentage**	**Frequency**	**Percentage**	**Frequency**	**Percentage**	
4	6	50.0	3	25.0	3	25.0	**100.0**
3	3	75.0	0	0.0	1	25.0	**100.0**
2	76	90.4	8	9.5	0	0.0	**99.9**
Total Frequecny	**85**		**11**		**4**		

Table 5. Changes in Frequency and Percentage of Tool Assemblage of Shurmai (GnJm 1) through Time

Unit	Local material		Non local material		Exotic material		Total percentage
	Frequency	**Percentage**	**Frequency**	**Percentage**	**Frequency**	**Percentage**	
4	2	11.1	9	50.0	7	38.9	**100.0**
3	4	50.0	2	25.0	2	25.0	**100.0**
2	44	84.6	8	15.4	0	0.0	**100.0**
Total frequency	**50**		**19**		**9**		

Raw Materials Utilized at the Kakwa Lelash Rockshelter (GnJm 2). Five stratigraphic units were identified in the Kakwa Lelash stratigraphy. Although most lithic material was recovered from Unit 3, the general pattern of utilization of lithic raw materials from each unit of the site are almost identical. This trend contrasts with that of Shurmai Rockshelter which evidently shows the change in lithic raw material procurement strategies through time. Analysis of the stratigraphy of Kakwa Lelash Rockshelter is still underway and a full cross dating of the Kakwa Lelash Rockshelter with the Shurmai Rockshelter is not yet complete. Aside from the occasional object of possible MSA affiliation, evidence of an occupation earlier than the Later Stone Age was not found in Kakwa Lelash Rockshelter *sondage*.

Local materials comprise 48.8 percent of the debitage assemblage from the Kakwa Lelash Rockshelter. The dominant material is milky or white quartz crystals (45.5 percent) which is found at the site. The presence of quartz source areas at the site is described in Chapter III. The percentage of basalt was only 3.1 percent at Kakwa Lelash, while at the Shurmai Rockshelter basalt was the primary local raw material. The quartz material found at the site is very tough and varies from mediocre to poor quality for knapping. This material is strong enough to be used as hard hammer stones.

Non-local materials are a very significant component of debitage assemblage from the Kakwa Lelash Rockshelter. Non-local materials comprise 42.0 percent of the total unmodified flake assemblage. When the percentage of utilized local (48.4 percent) and non-local (42.0 percent) materials of the unmodified flake assemblage is compared, the difference between these two materials is not significant. Exotic materials contribute only 9.2 percent to the debitage assemblage. Compared to the Shurmai Rockshelter, the utilization of non-local materials at the Kakwa Lelash Rockshelter is significantly higher and the utilization of exotic material is almost same. Local materials make up fully 75.0 percent of the core assemblage at the site. Non-local material contributes only 25.0 percent of the core assemblage. However, the dominant material found in core assemblage is also quartz (59.4 percent). No exotic core materials were recovered. These results may be due to the fact that fragments, chunks, and cores of less than 2 cm were not counted. In addition, many obsidian flakes carry characteristics of bipolar percussion technique. These flakes are very small, generally almost 1 cm. If the cores of these exotic flakes were also very small, they might not have been counted. However, this analysis discrepancy may not be significant. If the exotic materials were transported into the site as preformed blanks, the presence of exotic cores would be very low. In addition, if the exotic materials were heavily utilized due to the scarcity of the exotic materials near the site, most of exotic core materials became very small and were not counted because of their small size. This hypothesis will be tested later section.

Tool assemblage of Kakwa Lelash Rockshelter consists of 42.3 percent of local, 26.9 percent of non-local, and 30.8 percent of exotic materials. Compared to the Shurmai Rockshelter, the utilization of local materials decreased, while that of exotic material increased. The utilization of non-local materials shows little change. If the core analysis results are correct, this tool analysis result is not surprising. Based on debitage, core, and tool assemblage of Kakwa Lelash Rockshelter, both local and non-local materials were equally utilized for manufacturing stone artifacts. However, exotic material was still not significantly utilized in the site (Table 6).

In terms of units, the Unit 1 debitage assemblage consists of 50.0 percent of local and 50.0 percent of non-local materials. Local material utilized is quartz and non-local material is chert and rhyolite. In this unit, both local and non-local materials were equally utilized. However, the Unit 1 contributes only 0.6 percent of total debitage assemblage of the site (Table 7).

Local materials comprise 46.7 percent of the assemblage of the Unit 2, while non-local materials contribute 50.6 percent of the assemblage of the unit. The percentage of

Table 6. Frequency and Percentage of Lithic Assemblage of Kakwa Lelash (GnJm 2)

Material Type	Unmodified Flakes		Cores		Tools	
	Frequency	**Percentage**	**Frequency**	**Percentage**	**Frequency**	**Percentage**
Local material	520	48.8	24	75.0	11	42.3
Non local material	449	42.0	8	25.0	7	26.9
Exotic material	98	9.2	0	0.0	8	30.8
Total	**1067**	**100.0**	**32**	**100.0**	**26**	**100.0**

Table 7. Changes in Frequency and Percentage of Lithic Debitage Assemblage of Kakwa Lelash (GnJm 2) through Time

Unit	Local material		Non local material		Exotic material		Total percentage
	Frequency	**Percentage**	**Frequency**	**Percentage**	**Frequency**	**Percentage**	
5	13	54.2	8	33.3	3	12.5	**100.0**
4	176	48.5	139	38.3	48	13.2	**100.0**
3	184	48.4	154	40.9	40	10.6	**99.9**
2	122	46.7	132	50.6	7	2.7	**100.0**
1	3	50.0	3	50.0	0	0.0	**100.0**
Total Frequency	**498**		**436**		**98**		

exotic material in this unit is 2.7. In this unit, non-local materials appear to have been slightly preferred to local materials although the percentage differences between them is not significant.

The Unit 3 debitage assemblage consists of 48.4 percent local, 40.9 percent non-local, and 10.6 percent exotic materials. Although there is a small increase in the percentage of exotic material relative to Unit 2, both local and non-local materials were equally utilized in manufacturing stone artifacts.

The Unit 4 assemblage is composed of 48.5 percent local, 38.3 percent non-local, and 13.2 percent exotic material. Compared to the Unit 3, the percentage of non-local material decreased while that of exotic material increased. The percentage of local material did not change. Stone artifacts discovered from both Units 3 and 4 assemblages contribute most of the debitage in the assemblage from the site.

Local material comprises 54.2 percent of the Unit 5 assemblage, non-local 33.3 percent. Exotic material contributes 12.5 percent. Compared to the previous unit, both non-local and exotic materials appear to have been slightly less utilized, while local materials were slightly more utilized. Again, this change does not appear to be significant. In addition, the Unit 5 contributes only 1.3 percent of the total debitage in the assemblage at the Kakwa Lelash Rockshelter. In this analysis, 37 flakes recovered in cleaning the north wall were not included in this analysis because they lacked precise provenience.

In regard to the cores, local materials comprise 50.0 percent of the Unit 2 core assemblage. The percentage of non-local material is also 50.0 percent. Like the debitage assemblage, most of cores recovered from the site come from Unit 3 and Unit 4. Unit 3 contributes 34.4 percent and the Unit 4 contributes 31.3 percent of the total core assemblage. Local materials comprise 78.6 percent of the Unit 3 core assemblage. The dominant raw material is quartz. The Unit 4 core assemblage consists of 83.3

percent of local and 16.7 percent of non-local materials. Neither Unit 1 nor Unit 5 yielded any cores. This result suggests that local materials were preferred to non-local materials. However, as mentioned above, this result may actually indicate that non-local materials were transported into the site only after they had been "roughed out" and the cortex removed and discarded somewhere else (Table 8).

Only Unit 3 and Unit 4 produced tools (Table 9). Unit 3 consists of 40.0 percent local, 26.7 percent of non-local, and 33.3 percent of exotic materials. Local materials comprise 45.5 percent of the Unit 4 tool assemblage, while non-local materials contribute 27.3 percent of the assemblage. Exotic material comprises 27.3 percent of the assemblage. Compared to core analysis of the site, the percentage of both exotic and non-local materials are significant. Again, these percentages suggest that cores of both exotic and non-local materials must have been roughed out somewhere else before they were transported into the site.

Transportation of Raw Materials into the Sites. To test the hypothesis suggested above, unmodified flakes were grouped into primary, secondary, and tertiary types. The definitions of each of these flake types are described in the methodology chapter. This test is useful for analyzing how particular raw materials were transported into the sites. That is, this analysis can provides us information about whether a given material was imported as raw nodules, roughed out cores, or tool blanks. It is assumed that if raw materials were procured at a distance from the site, the cortex on the raw materials would have been removed somewhere else before they were transported to the site in order to minimize the transport cost. If the raw material was of good quality for flaking and rarely found at or near the site, the raw material would have been roughed out somewhere else in order to maximize the useable material and reduce its weight.

Shurmai Rockshelter (GnJm 1). Tertiary flakes comprise 64.1 percent of local materials recovered at this site. Secondary flakes contribute 28.6 percent and primary flakes 7.3 percent of local materials. Flakes made of non-local materials consist of 70.4 percent of tertiary, 22.5

percent of secondary, and 7.1 percent of primary flakes. In the case of exotic material, tertiary flakes comprise 88.2 percent of the debitage, secondary flakes 9.8 percent and primary flakes 2.0 percent of it (Table 10).

The majority of flakes recovered from the site are tertiary flakes. This indicates that most of the raw materials transported to the site arrived there after they had been roughed out at somewhere else. The percentage of non-local and exotic tertiary flakes is very high, while the percentage of primary flakes is very low. Therefore, there is strong correlation between type of raw material and type of flakes. It appears that non-local and exotic materials arrived at the site as tool blanks or almost fully decorticated cores. We can assume that the occupants of the site minimized their transport cost of the non-local and exotic materials.

Kakwa Lelash Rockshelter (GnJm 2). Local materials consist of 92.9 percent of tertiary, 4.4 percent of secondary, and 2.7 percent of primary flakes. Flakes made of non-local materials consist of 80.8 percent tertiary, 14.8 percent secondary, and 2.2 percent primary flakes. The majority of the exotic materials are tertiary flakes (96.0 percent) while the percentage of primary flakes of exotic material is only 1.0 (Table 11).

Based on the analysis result, the percentage of tertiary flakes recovered from the Kakwa Lelash Rockshelter is more than 80.0 percent of all material types. Compared to Shurmai Rockshelter, the percentage of primary and secondary flakes recovered from Kakwa Lelash is significantly lower. That is, the majority of the raw materials arrived at Kakwa Lelash after it had been roughed out or severely decorticated somewhere else. Probably much of the material was transported to the site as tool blanks.

In summary, the raw material analyses of the two archaeological sites, Shurmai (GnJm 1) and Kakwa Lelash (GnJm 2) rockshelters reveal that the occupants of the two sites practiced very different raw material procurement strategies. Further, the raw material procurement strategies practiced at the Shurmai site changed gradually through

Table 8. Changes in Frequency and Percentage of the Core Assemblage of Kakwa Lelash (GnJm 2) through Time

Unit	Local material		Non local material		Exotic material		Total Percentage
	Frequency	**Percentage**	**Frequency**	**Percentage**	**Frequency**	**Percentage**	
4	10	83.3	2	16.7	0	0.0	**100.0**
3	11	78.6	3	21.4	0	0.0	**100.0**
2	3	50.0	3	50.0	0	0.0	**100.0**
Total Frequency	24		8		0		

Table 9. Changes in Frequency and Percentage of the Tool Assemblage of Kakwa Lelash (GnJm 2) through Time

Unit	Local material		Non local material		Exotic material		Total Percentage
	Frequency	**Percentage**	**Frequency**	**Percentage**	**Frequency**	**Percentage**	
4	5	45.5	3	27.3	3	27.3	**100.1**
3	6	40	4	26.7	5	33.3	**100.0**
Total frequency	**11**		**7**		**8**		

Table 10. Frequency and Percentage of Debitage Type of Shurmai Rockshelter (GnJm 1)

Flake type	Local material		Non local material		Exotic material	
	Frequency	**Percentage**	**Frequency**	**Percentage**	**Frequency**	**Percentage**
Primary	81	7.3	22	7.1	1	2.0
Secondary	316	28.6	70	22.5	5	9.8
Tertiary	708	64.1	219	70.4	45	88.2
Total	**1105**	**100.0**	**311**	**100.0**	**51**	**100.0**

Table 11. Frequency and Percentage of Debitage Type of Kakwa Lelash Rockshelter (GnJm 2)

Flake type	Local material		Non local material		Exotic material	
	Frequency	**Percentage**	**Frequency**	**Percentage**	**Frequency**	**Percentage**
Primary	14	2.7	10	2.2	1	1.0
Secondary	23	4.4	68	14.8	3	3.1
Tertiary	483	92.9	371	80.8	94	96.0
Total	**520**	**100.0**	**459**	**97.8**	**98**	**100.1**

time. The oldest unit, Unit 2, which belongs to Middle Stone Age, reveals that occupants of the site were heavily dependent on local material. Unit 3, which belongs to the later Middle Stone Age, shows a change in these procurement strategies. In this unit, local materials were less utilized, while the utilization of exotic material increased. Unit 4, which belongs to the Later Stone Age, reveals that the inhabitants of the site depended on non-local and exotic materials more than local materials. In this unit, non-local and exotic materials comprise more than 80.0 percent of the assemblage.

Kakwa Lelash (GnJm 2) Rockshelter, which dates to the Later Stone Age, does not show much change in raw

material procurement strategies. The occupants at this site practiced non-localized raw material procurement strategies. Although they still utilized many local materials, the preference of local material was quartz, not basalt, which were heavily used in Shurmai Rockshelter. As a result, chi-square of the raw material types utilized for manufacturing stone artifacts in two sites is very significant (Table 12).

The result of both the core and the tool analyses of the Shurmai assemblage are consistent with the results of the analyses of the debitage assemblage (Figure 13). Most cores and tools were made of local materials. The numbers of non-local and exotic cores were not significant although

Table 12. Chi-square of Raw Material Types of the Two Archaeological Sites, Shurmai (GnJm 1) and Kakwa Lelash (GnJm 2).

Material type	GnJm 1	Gn Jm 2	Total
[a] Basalt	1000 96.8 68.2	33 3.2 3.1	1033 40.8
[a] Quartz	103 17.5 7.0	485 82.5 45.5	588 23.2
[a] Quartzite	1 33.3 .1	2 66.7 .2	3 .1
[a] Schist	1 100.0 .1	 .0 .0	1 .0
[b] Chert	235 34.4 16.0	448 65.6 42.0	683 27.0
[b] Rhyolite	73 100.0 5.0	 .0 .0	73 2.9
[b] Sandstone	2 66.7 .1	1 33.3 .1	3 .1
[b] Tuff	1 100.0 .1	 .0 .0	1 .0
[c] Obsidian	51 34.2 3.5	98 65.8 9.2	149 5.9
Column total	1467 57.9	1067 42.1	2534 100.0

Chi-square	Value	D.F.	Significance
Pearson	1279.0344	8	.0000
Likelihood ratio	1533.185	8	.0000

[a] : local material   [b]: non local material   [c]: exotic material

the numbers of cores and tools made of these materials showed some increase through time.

At Kakwa Lelash (GnJm 2), core and tool analyses returned quite interesting results. In spite of non-localized raw material procurement strategies, this site produced many cores that were made of local materials. However, tools made of non-local and exotic materials were more abundant than those of local materials. It was hypothesized, therefore, that most of non-local and exotic

materials were transported to the site after it had been roughed out or decorticated somewhere else.

This hypothesis is strongly supported by the debitage type analysis. The debitage type analysis at both sites returned the same result. Regardless the type of raw material, tertiary flakes comprise more than 60.0 percent of debitage assemblage. Comparing the two sites, Kakwa Lelash Rockshelter produced more tertiary flakes than Shurmai Rockshelter. As a result, both secondary and primary

Non-local

②	GnJm1-2
③	GnJm1-3
④	GnJm1-4
❷	GnJm2-2
❸	GnJm2-3
❹	GnJm2-4

Local   Exotic

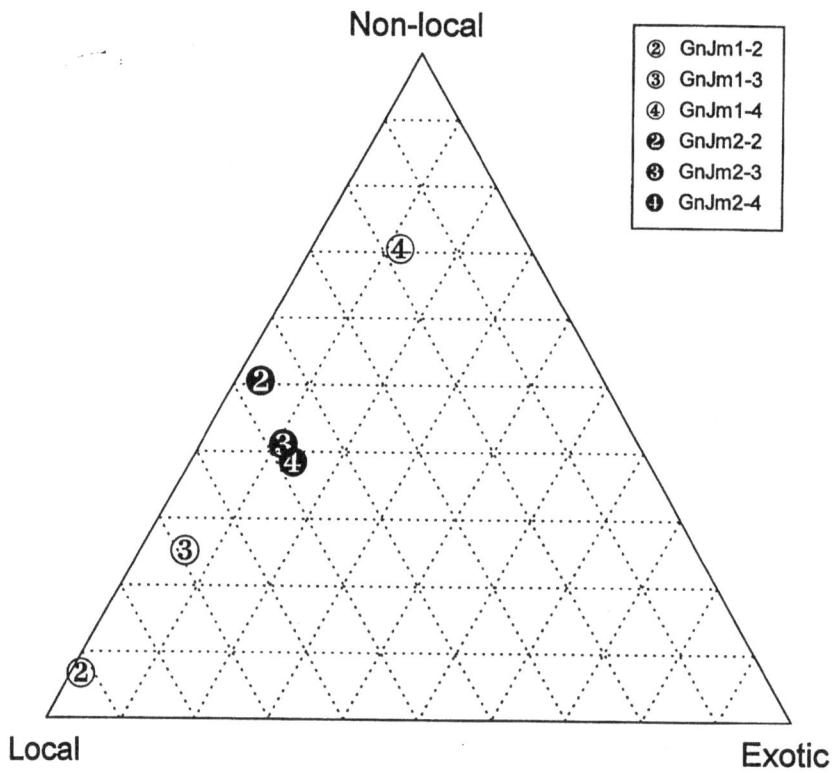

Figure 13. Variability in Raw Material Procurement of the Two Sites.

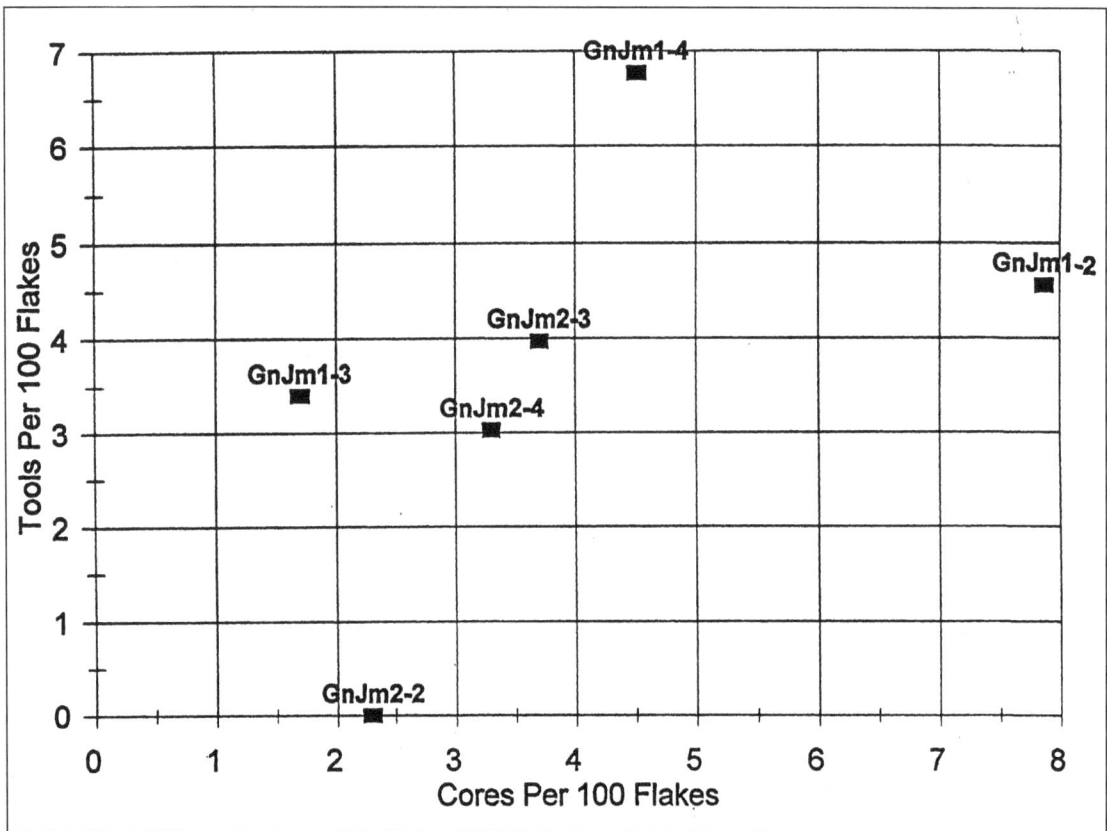

Figure 14. The Ratio of Cores to 100 Flakes and the Ratio of Tools to 100 Flakes of the Two Sites.

flakes contribute less to its assemblage. More tertiary flakes and less primary and secondary flakes indicates that raw materials that arrived at the Kakwa Lelash had been severely reduced or decorticated before their transportation. That is, they were transported into the site as tool blanks.

In the relative ratio of core and tool case, non-local and exotic material produced high frequency of tools relative to the frequency of cores at both sites. Correspondingly, local materials produced low frequency of tools relative to their core frequency. This probably reflects the quality of stone available for the knapping rather than the skill of the knappers. That is, the dominant local materials, basalt and quartz, are mediocre to poor for knapping, while the non-local and exotic materials, such as chert and obsidian, are good for knapping and make fine tools.

In terms of lithic variability, the ratio of cores to 100 flakes and the ratio of tools to 100 flakes were analyzed in order to understand how many cores and tools were produced relative to the number of flakes (Figure 14). Shurmai Rockshelter shows more lithic variability in cores, while Kakwa Lelash Rockshelter shows more in tools. This is because Shurmai site produced more varied types of core materials than Kakwa Lelash Rockshelter. No exotic core material was found in the latter site. In contrast, at the latter site raw material types used to produce tools were more various than those of cores. Figure 14 indicates that vertical line is related to tool material procurement and horizontal line to core material procurement strategies. Figure 14 suggests that residents of the Shurmai Rockshelter practiced a planned raw material procurement strategy because the raw materials are more horizontally distributed. In contrast, Kakwa Lelash Rockshelter residents had an embedded raw material procurement strategy as raw materials there are more vertically distributed. However, as analyses of raw materials, flake types, and cores revealed, most of the raw materials transported to the Kakwa Lelash Rockshelter are non-local materials. In addition, the lithic industry of the Kakwa Lelash Rockshelter is characterized by microlithics. In the latter case, the cores are maximized and exhausted. As a result, very small cores and chunks resulting from the poor raw material quality were not counted. Therefore, we must not interpret raw material procurement strategies practiced at two rockshelters on the basis of the lithic variability that appears in the Figure 14.

Techno-Morphological Analysis

One of the important variables for understanding morphological change in stone artifacts is size. Size can be understood in terms of three dimensions, length, width, and thickness. In order to understand this morphological change at the two sites, the size of whole, unmodified flakes and their exterior striking platform angles were measured. The relationship between size and exterior angle can also provides information about technological change.

Debitage Analysis

Morphological analysis of debitage assemblages from Shurmai (GnJm 1) and Kakwa Lelash (GnJm 2) rockshelters reveal interesting results (Table 13). In the Table 19, the number of each Unit represents different time period. Unit 2 of the Shurmai dates to the Middle Stone Age, Unit 3 to the Later Stone Age, while all units from Kakwa Lelash date to the Later Stone Age. At Shurmai Rockshelter, morphological change is quite distinct. Through time, the average length, width, and thickness noticeably decreased.

Table 13. Morphological Changes through Time

Site	Unit	Average length (mm)	Average Width (mm)	Average Thickness (mm)	Average Angle of Platform
Shurmai (GnJm 1)	4	29.37	21.18	6.29	58.3
	3	34.53	27.80	8.04	60.7
	2	44.72	38.34	10.98	65.4
Kakwa Lelash (GnJm 2)	5	21.25	18.94	6.02	69.0
	4	22.51	19.98	6.37	72.54
	3	23.14	20.51	6.62	74.13
	2	23.74	20.68	6.45	73.72
	1	25.83	20.20	6.77	62.50

Although the morphological change in flakes from Kakwa Lelash Rockshelter is not as distinguishable, a gradual decrease in flake size through time did occur. Compared to the Shurmai Rockshelter, Kakwa Lelash Rockshelter produced smaller flakes. This result supports my reconstruction of the raw material procurement strategies at the two sites. That is, raw materials arrived at Kakwa Lelash site as tool blanks that produced abundant tertiary flakes in the site. Shurmai site also produced more tertiary flakes than secondary and primary flakes.

The major morphological change from the Middle to the Later Stone Age in East Africa is the decrease in flake size. This is particularly evident in the later Middle Stone Age due to the appearance of blade production. In order to understand this morphological change, the flake type was analyzed and flake size was measured. In general, flakes recovered from the Later Stone Age units at Kakwa Lelash Rockshelter are smaller than those from the Middle Stone Age units at Shurmai Rockshelter in terms of length, width, and thickness. Flakes recovered from the two archaeological sites strongly support this general trend towards decrease in flake size through time. Table 14 shows the changes in size. Because there is no significant difference between values of entire flake assemblage (both complete and broken flakes) and only complete flake assemblage, the t-test of entire flake assemblage is presented.

Another characteristics of the later Middle Stone Age is an appearance of blades which are utilized as composite tools in the Later Stone Age. Shurmai Rockshelter produced 139 blades, Kakwa Lelash 36 blades. At the Shurmai Rockshelter, the debitage assemblage consists of 843 complete and 624 broken flakes. Thus, blades comprise 16.49 percent of the complete flakes there. Contrary to my expectation, the debitage assemblage recovered from Kakwa Lelash consists of 587 complete and 480 broken flakes. Of these, blades make up only 6.13 percent of total number of complete flakes. Thus, Shurmai produced more blades. However, when debris is considered, Kakwa Lelash Rockshelter produced four times as many blades as Shurmai. This is because flakes recovered from Kakwa Lelash Rockshelter are more fragile than those from Shurmai, as the flakes from the former are smaller and thinner than those from the latter.

Flake termination and flake butt types were analyzed because these variables are closely related to each other and reveal the technical skills of their makers. As described in the Chapter V, feathered termination can be affected by how much a knapper controls his skill. Feathered termination is generally produced by low exterior platform angle with constant or confined platform thickness. If the constant platform thickness is variable, high platform angle result in feathered termination and

longer flakes. In general, correct flaking and high exterior platform angle with variable platform thickness result in feathered termination. That is, combination of more technologically skillful method leads to high production of feathered termination.

The Shurmai Rockshelter produced 757 complete flakes with feathered termination, while Kakwa Lelash 532 feathered flakes. That is, flakes with feather termination comprise 89.8 percent of the complete flake assemblage at Shurmai Rockshelter and 90.6 percent of those of Kakwa Lelash. Therefore, the difference between two sites on this dimension is not significant. Rather they are almost identical.

In terms of flake butt type or striking platform of the flake, if a knapper takes care of preparing striking platform on the core more intensely to produce desirable flakes, the butt of flakes detached from the prepared core is characterized by faceted or retouched platform. When the striking platform is retouched or faceted, the exterior platform angle is adjusted and thereby the distribution of force in flake removal is controlled. As a result, the knapper can produce more desirable flakes. When the two archaeological sites are compared, the Shurmai Rockshelter is found to have 626 flakes with retouched butt or striking platform. These flakes comprise 42.6 percent of flake assemblage. Flakes with retouched butt contribute 41.8 percent (N ' 446) of flake assemblage of Kakwa Lelash Rockshelter. Although the difference between these two site are not great, the Later Stone Age units at Kakwa Lelash Rockshelter produced fewer retouched platforms than Shurmai. This issue will be discussed in the section on cores below. In spite of this slight decrease in retouched platform, in general, the technological differences between these sites are not noticeable. At best, the knappers at Kakwa Lelash Rockshelter may be said to have practiced a little more refined technology than those at Shurmai Rockshelter.

Core Reduction Sequence and Production of Tool Blanks

The Nature of Cores. In order to reconstruct the core reduction sequence at the two sites, it is necessary to understand the nature of cores discovered from the sites and production of the tool blank, which results from the preparation of decortication. In this section I will discuss the nature of the cores recovered at the two archaeological sites together with the nature of the tool blank from them. The cores discovered from the two archaeological sites are classed as either complete or fragmentary cores. Only complete cores are considered in this section. Because the core fragments can not provide the whole patterns of flake scars left on the cores or information about the size and nature of the cores, fragmentary cores were excluded from the core analysis.

Table 14. Comparison of Complete Flake Size

Dimension	Shurmai (GnJm 1) (N=1467)				Kakwa Lelash (GnJm 2)(N=1067)			
	Mean	**(S. d.)**	**Median**	**Range**	**Mean**	**(S.d.)**	**Median**	**Range**
Length (mm)	41.12	16.67	38.3	5.90 - 123.90	22.97	7.35	21.7	7.90 - 58.10
Width (mm)	34.85	21.84	32.8	3.30 - 675.20	20.24	7.06	19.0	1.50 - 62.20
Thickness (mm)	10.03	4.82	9.3	1.50 - 63.60	6.41	2.62	5.9	2.10 - 24.20

Shurmai (GnJm 1) flakes vs. Kakwa Lelash (GnJm 2) flakes

Length: t = 37.04   df = 2143.53   2-tail prob. = .000   Mean difference = 18.147
Levene's test of homegeneity of variance: F = 553.645   P = .000

Width: t = 23.95   df = 1862.91   2-tail prob. = .000   Mean difference = 14.607
Levene's test of homegeneity of variance: F = 120.318   P = .000

Thickness: t = 24.29   df = 2361.05   2-tail prob. = .000   Mean difference = 3.622
Levene's test of homegeneity of variance: F = 209.518   P = .000

Shurmai Rockshelter (GnJm 1). Complete cores comprise 54.0 (N ' 54) percent of Shurmai core assemblage and core fragments 46.0 percent. In order to understand how many platforms on the core were prepared to produce flakes, complete cores are subdivided into unidirectional, bi-directional, radial, random, and undetermined cores based on the flake scars left on the cores. When cores show unidirectional flake scars, it means that only a single platform was prepared on the core. These types of cores are, in general, related to blade production. If cores carry bi-directional flake scars, it indicates cores with two platforms were prepared with one opposite the other. When cores were prepared for multiple platforms, they show generally radial scar pattern on them. However, we cannot be certain whether random flake scar patters resulted from intention or accident. Complete cores discovered from Shurmai site consist of 85.2 percent (N ' 46) radial, 7.4 percent (N ' 4) random, 5.6 percent (N ' 3) unidirectional, and 1.9 percent (N ' 1) undermined cores. The Shurmai Rockshelter did not yield any bi-directional cores; multiplatform cores dominated the assemblage (Table 15).

Of radial cores, at least 10 of 54 cores show flaking on one surface only. On these cores, the cortex is presented in the center of the opposite unflaked surface. However, flaked surfaces of these cores do not show large noticeable negative bulb of scar which is characteristic of Levallois method. As a result, these flakes were categorized as discoidal cores rather than Levallois cores. Still, we cannot exclude the possibility of the presence of Levallois cores

because almost all cores recovered from Shurmai site are so severely weathered that it is very hard to read the limit of flake scars. In some cases, the cores that are categorized as whole or complete show some broken part on them or some new scars. In addition, 84.8 percent (N ' 39) of core fragments are made of basalt which are severely weathered and are hard to produce desirable flakes. These fragment cores cannot provide whole information about the nature of core as mentioned above. That is, we cannot be sure whether the absence of true Levallois core in the Shurmai Rockshelter was caused by the limitations imposed by raw materials or was due to ignorance of the Levallois technique as the part of the knappers.

The complete cores discovered from the site consisted of exhausted and partially used types (Table 16). Partially used cores are characterized by many step and hinge fractures. These fractures would be caused by the nature of the basalt raw material of many of these cores.

As these cores were discarded before their exhaustion, they tend to be larger than those of other materials. However, even exhausted cores of basalt are generally large enough to produce more flakes using different techniques. Since most of cores (N ' 75 of 100) of the Shurmai Rockshelter are basalt, the size of these cores is, in general, large. Comparing length to width, the mean value of the width is greater than that of length. In order to characterize the shape of cores, two indices were used: length by width and width by thickness. In general, when the value of the length by width is greater than that of width by thickness, the shape of cores is elongated. When the values show the opposite ratio, the shape is generally short and broad. Many of cores recovered from the Shurmai Rockshelter are shorter and more square in shape rather than elongated or round.

Kakwa Lelash (GnJm 2) Rockshelter. The Kakwa Lelash Rockshelter produced 25 complete cores (78.1 percent) and 7 core fragments (21.9 percent). Of the complete cores, those with a radial scar pattern comprise 76.0 percent of complete core assemblage. Both cores with bi-directional and random flake scar pattern contribute 12.0 percent of the complete core assemblage. Unlike the Shurmai Rockshelter, the Kakwa Lelash Rockshelter did not produce cores with unidirectional flake scar patterns. Instead, this site produced cores with bi-directional flake scar pattern. Cores exhibiting a radial scar pattern were dominant at both sites. One interesting result of this analysis is that although blades were recovered from both sites, neither produced definite blade cores. Of course the exhausted cores did not reveal whether they were used for the production of blade or not in their early stages before their exhaustion. Further, some fragment cores and exhausted cores show that radial scar patterns on their face developed toward their side rather

than their flat surfaces. It is possible that these cores could have been used to produce blades.

In terms of size, the cores recovered from Shurmai Rockshelter are bigger than those from Kakwa Lelash Rockshelter. This decrease in core size was probably caused either to efforts at maximizing the use of raw material or by the general tendency of Later Stone Age knappers in Africa to decrease flake size.

Production of Tool Blanks. In the lithic reduction process, as in the production of tool blanks, the first task is to produce desirable flakes. Thus, knowledge of the tool blank can provide us with information about the first step in the core reduction sequence. In order to reconstruct the core reduction sequence, only complete flake tools and complete unmodified flakes were analyzed.

Complete Flake Tools and Unmodified Flakes. Tools discovered from the two rockshelters were divided into cobble, flake, and non-flaked or ground tools. Shurmai Rockshelter produced three cobbles, 32 complete flakes, 41 incomplete flakes, and three non-flaked tools. Kakwa Lelash Rockshelter yielded seven complete flakes, 16 incomplete flakes, and three non-flake tools.

Table 15. Core Completeness and Flake Scar Pattern on Cores

Completeness		Shurmai (GnJm 1)		Kakwa Lelash (GnJm 2)	
		Frequency	**Percentage**	**Frequency**	**Percentage**
	Complete	54	54.0	25	78.1
	Fragment	46	46.0	7	21.9
Flake Scar Pattern	Unidirection	3	5.6	0	0.0
	Bidirection	0	0.0	3	12.0
	Radial	46	85.2	19	76.0
	Random	4	7.4	3	12.0
	Undetermined	1	1.9	0	0.0

Table 16. Size of Complete Cores

Size (mm)	Shurmai (GnJm 1)				Kakwa Lelash (GnJm 2)			
	Mean	**(s.d)**	**Median**	**Range**	**Mean**	**(s.d)**	**Median**	**Range**
length	55.1	12.0	27.4	25.5 - 87.6	41.1	15.9	32.8	22.0 - 83.4
Width	557.6	11.7	56.8	27.6 - 93.9	47.7	18.3	34.4	19.9 - 81.0
Thickness	28.3	10.6	27.4	8.8 - 55.0	28.6	15.7	25.5	9.6 - 27.3
Length/width	9.7	.16	.94	.7 - 1.3	.94	.17	.91	.7 - 1.3
Width/thickness	2.29	.80	2.2	1.1 - 4.2	1.74	.53	1.63	1.1 - 3.3

Shurmai (GnJm 1) Rockshelter. Complete flake tools comprised 40.5 percent of the tool assemblage at the site. These complete flake tools consist of 21.9 percent (N ' 7) secondary flakes and 78.1 percent (N ' 25) tertiary flakes. Interestingly, all seven secondary flakes are naturally backed flakes (Figure 15).

Unmodified flakes recovered from the Shurmai Rockshelter consist of 843 (57.5 percent) complete and 624 (42.5 percent) broken flakes. Of 843 unmodified complete flakes, primary flakes comprise 7.4 percent of the assemblage. Secondary flakes contribute 32.5 percent of the assemblage, while tertiary flakes comprise 60.1 percent of it. As in the flake tool assemblage, tertiary flakes are dominant in the complete flake assemblage. It appears that knappers at the site thought tertiary flakes produced the most desirable flake tools. All secondary flakes in the tool assemblage are naturally backed flakes. The unmodified complete flake assemblage also reveals that naturally backed flakes comprise 28.1 percent of secondary flakes. These results do not seem to be accidental. Rather they indicate that naturally backed flakes were preferred by knappers as tools and that the knappers were trying to produce either tertiary flakes or naturally backed flakes as desirable end products. One way of understanding platform preparation on a core is to examine flake scars. These scars are remnants of core preparation and carry information about flaking platforms. For instance, a single platform core was used to produce blades. assemblage. Flakes with unidirectional flake scar patterns comprise 41.7 percent of flake assemblage. Another method for producing flakes is to use multiple platforms on a core. The multiple platform method is reflected on flakes with radial flake scar patterns. This method comprises 26.1 percent of flake assemblage. Flakes with bi-directional or random direction platforms contribute only 3.5 percent and 8.2 percent of flake assemblage respectively. Unfortunately, 128 flakes (20.5 percent) did not reveal their flake scar pattern.

Since both unidirectional and multidirectional platform preparation are systematic methods designed to get as many flakes as much as possible, this analysis indicates that knappers of the site maximized raw material to produce desirable flakes. Both flake tool assemblage and unmodified complete flake assemblage are dominated by single platform. That is, the single platform on a core would be preferred to get desirable end product. Furthermore, it can be assumed that many unmodified complete flakes were cached for the future rather than accidental byproduct produced during the core reduction process. Although this result is not strongly correlated with the majority of core types, which carry radial flake scars and are characterized by multiplatforms, flakes with radial flake scar patterns comprise significant portion of the total percentage. However, this discrepancy between dominant platform type and dominant flake scar pattern on cores may reflect that the platform types on cores were

Correspondingly, most blades carry flake scars that are unidirectionally parallel. That is, blade scars show how the core platforms were prepared.

Thirty-two complete flake tools recovered from Shurmai Rockshelter were divided into several groups on the basis of number of their flake scars. After this subdivision, only flakes tools that have more than three flake scars were analyzed. This is because flake tools with more than three flake scars can provide certain flake scar direction without ambiguity. Of 32 flake tools, 19 flake tools carried more than three flake scars. Unfortunately, I was not able to read flake scars on complete flake tools because these tools were too severely weathered to show the direction of flake scars. Of these 19 tools, seven flakes did not reveal the direction of flake scar. The rest of flake tools show the various directions of the flake scars were either unidirectional (N ' 5), bi-directional (N ' 2), radial (N ' 2), or random (N ' 2). As unidirectional flake scars are dominant, single platform on a core would be preferred to produce desirable flakes. There are no differences among bi-directional, radial, and random flake scar pattern. That is, cores with two platforms opposite each other and multiplatforms would be made to produce tool blank although cores with single platform would be preferred.

Unmodified complete flakes with more than three flake scars comprise 74.0 percent (N ' 624) of the flake

changed through the reduction process in order to maximize the raw materials. This is indicated by most of cores recovered from the site. They are dominated by exhausted cores (Table 17).

Kakwa Lelash (GnJm 2) Rockshelter. Only seven complete flake tools were recovered from Kakwa Lelash Rockshelter. None of these were made on primary or secondary flakes. All these seven tools were made on tertiary flakes (100.0 percent) (Table 18).

Unmodified complete flakes discovered from the site contribute 55.0 percent (N ' 587) of the flake assemblage, while broken flakes made up 45.0 percent (N ' 480). Complete flakes consist of 1.9 percent (N ' 11) primary, 11.2 percent (N ' 66) secondary, and 86.9 percent (N ' 510) tertiary flakes. Tertiary flakes are dominant in both the flake tool and unmodified complete flake assemblages. Compared to Shurmai Rockshelter, the tool assemblage at Kakwa Lelash Rockshelter contains fewer naturally backed flakes in secondary flake group. There is an apparent decline in naturally backed flakes in the tool assemblage (0.0 percent). Knappers of the Kakwa Lelash Rockshelter continued to try to produce tertiary flakes exclusively to get desirable flake tools, but they ceased making naturally backed flakes for tools. The decline in production of naturally backed flakes markedly distinguishes the Kakwa Lelash from the Shurmai lithic industry.

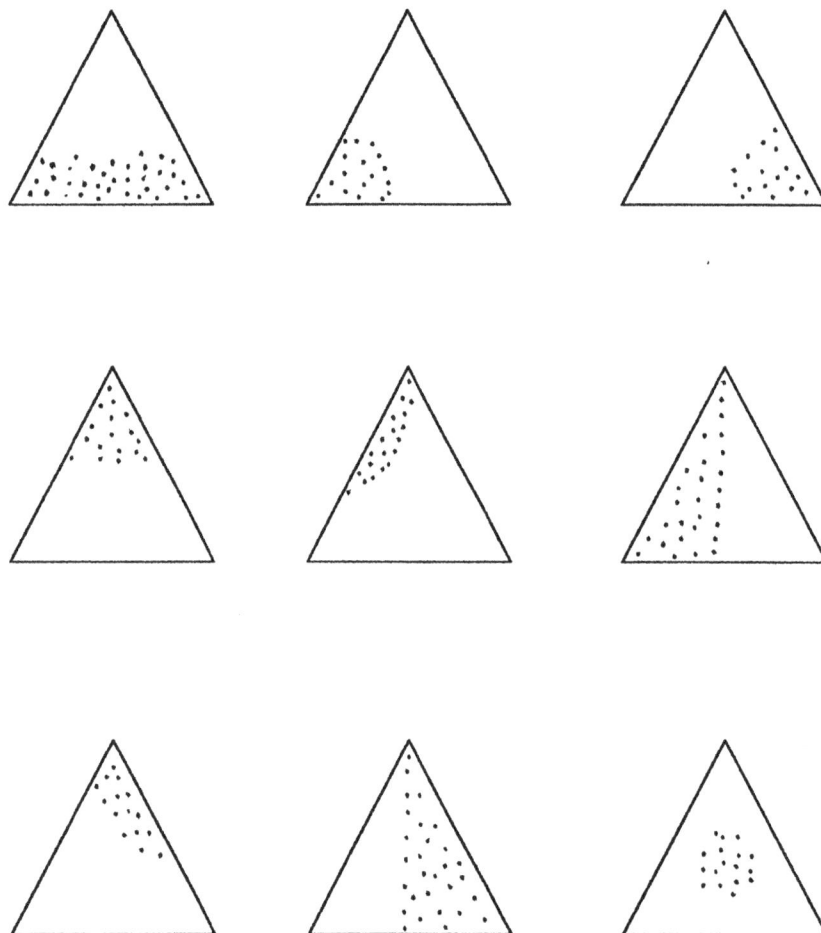

Figure 15. Positions of Cortex on the Exterior Surface of the Secondary Flakes.

Table 17. Complete Flake Tool and Unmodified Complete Flake Types and Flake Scar Patterns of Shurmai (GnJm 1) Rockshelter

Flake type		Flake tools		Unmodified flakes	
		Frequency	Percentage	Frequency	Percentage
	Primary	0	0.0	62	7.4
	Secondary	7	21.9	274	32.5
	Tertiary	25	78.1	507	60.1
Flake scar pattern	Unidirection	5	26.3	260	41.7
	Bidirection	3	15.8	22	3.5
	Radial	2	10.5	163	26.1
	Random	2	10.5	51	8.2
	Undetermined	7	36.8	128	20.5

Table 18. Complete Flake Tool and Unmodified Complete Flake Types and Scar Patterns of Kakwa Lelash (GnJm 2) Rockshelter

Flake type		Flake tools		Unmodified flakes	
		Frequency	Percentage	Frequency	Percentage
	Primary	0	0.0	11	1.9
	Secondary	0	0.0	66	11.2
	Tertiary	7	100.0	510	86.9
Flake scar pattern	Unidirection	3	60.0	208	55.6
	Bidirection	0	0.0	11	2.9
	Radial	1	20.0	71	19.0
	Random	0	0.0	28	7.5
	Undetermined	1	20.0	13	14.2

In terms of platform preparation on a core, five complete flake tools recovered from the Kakwa Lelash Rockshelter carry more than three flake scars on their exterior surface. These five tools are composed of three unidirectional, one radial, and one undetermined flake scars. The Kakwa Lelash Rockshelter produced 374 (63.7 percent) unmodified complete flakes which are characterized by than 3 flake scars on their exterior surface and 213 flakes (36.3 percent) with less than 3 flake scars. Flakes with more than three flake scars consist of 55.6 percent (N ' 208) unidirectional, 19.0 percent (N ' 71) of radial, 7.5 percent (N ' 28) random flake scar patters. Unfortunately, flake scar patterns of 14.2 percent (N ' 53) of complete flakes were not identified. Like flake tools, unidirectional flake scar pattern dominates the unmodified flake assemblage. These results indicate that at both sites, knappers preferred to prepare single platform on a core to get their desired end product. This is because that there are no significant differences between flake tools and unmodified flakes on the basis of types of platform preparation, flake scar patterns on them, and types of flakes. In addition, my experience at Dr. Callahan's flint knapping camp gave me some impression that whenever a knapper produces potential tool blanks or flakes large enough to be transformed into a tool during his core reduction process, he saves these blanks or flakes for the future.

In summary, knappers at both sites prepared cores with single platform and sought to produce many tertiary flakes as desired end products. Many unmodified complete flakes would be tool blanks stored for the future rather than byproducts produced during core reduction process or tool making. Although naturally backed flakes were favored as end product in early period, this preference disappeared through time. Unidirectional and radial flake scar patterns

are the dominant flake scar patterns found on the flakes recovered from both sites. This indicates that the knappers at the two sites used systematic flake removal methods in order to conserve raw materials. It also reflects use of a consistent methodology for the flake removal.

Platform Preparation and Flake Removal. As mention above, a prepared striking platform produces a more desirable flakes from a core because it allows the adjustment of exterior platform angle and gives the knapper greater the control of the distribution of force. Therefore, it is important understanding how platforms were prepared to produce desirable end product.

Shurmai Rockshelter (GnJm 1). Complete flake tools discovered from the site were subdivided into five platform types: cortical, plain, faceted, retouched, broken, and miscellaneous platforms. Plain platforms comprise 28.1 percent (N ' 9), faceted platforms 18.8 percent (N ' 6), retouched platforms 43.8 percent (N ' 14), broken platforms 5.3 percent (N ' 2), and miscellaneous 3.1 percent (N ' 1) of flake tool assemblage. There were no single cortical platforms in this assemblage. The dominant type is the retouched or faceted platform. Plain platforms are also significant in the assemblage. Thus, the end products were produced after decortication of a core without adjustment of platform or by adjusting the platforms. It appears that the knappers of the site preferred to adjust platforms in order to produce desirable end product.

The unmodified complete flake assemblage consists of 3.6 percent (N ' 30) cortical, 48.4 percent (N ' 408) plain, 21.2 percent (N ' 179) faceted, 19.5 percent (N ' 164) retouched, 6.6 percent (N ' 56) broken, and 0.6 percent (N ' 5) miscellaneous platforms. The dominant platform is the plain type. If the retouched platforms are included in faceted category, faceted platforms comprise 40.7 percent of complete flake assemblage. Therefore, there is a consistent pattern between flake tool assemblage and complete flake assemblage. This result suggests that many of the complete flakes were possible tools or tool blanks (Table 19).

In general, the faceting of the platform affects the exterior platform angle and the distribution of force in flake removal. In particular, the exterior platform angle together with the platform thickness greatly influences the size of the resulting flake and its attachment from a core. In general, the greater the exterior platform angle, the longer the flake. However, in this case the platform thickness becomes variable. All things being equal, the larger the platform thickness the longer and thicker the flake (Dibble

and Whittaker 1981; Speth 1972,1974). In order to understand how desirable flakes were removed from the core, the size of complete flake tool assemblage and unmodified complete flake assemblage was analyzed (Table 20). In terms of flake size, the size of flake tools is larger than that of unmodified flakes. To understand whether size is correlated with platform thickness and exterior platform angle in both assemblages, platform thickness and exterior platform angle of both flake assemblages were also measured.

According to this analysis, mean platform thickness of flake tools is larger than that of unmodified flakes. In addition, platform angle produces the same result. That is, thicker platform produce longer, wider, and thicker flakes. Thus, it appears that larger exterior platform angle leads to the production of longer flakes.

Table 19. Platform Preparation for Tool Blank of Shurmai (GnJm 1) Rockshelter

Platform type		Complete flake tools		Unmodified complete flakes	
		Frequency	**Percentage**	**Frequency**	**Percentage**
	Cortical	0	0.0	30	3.6
	Plain	9	28.1	408	48.4
	Faceted	6	18.8	179	21.2
	Retouched	14	43.8	164	19.5
	Broken	2	5.3	56	6.6
	Undetermined	1	3.1	5	0.6
Total		32	99.1	843	99.9

Table 20. Comparison of Flake Tool Size and Complete Flake Size of Shurmai (GnJm 1) Rockshelter

	Complete flake tools				Unmodified complete flakes			
	Mean	(S.d)	Median	Range	Mean	(S.d)	Median	Range
Length	53.0	22.2	53.8	3.0 - 109.0	44.4	16.6	42.6	5.9 - 123.9
Width	40.3	16.9	41.2	4.8 - 79.1	35.9	14.3	34.5	4.3 - 99.0
Thickness	12.1	5.4	12.7	4.1 - 25.9	10.5	4.4	10.0	1.5 - 33.2
Platform thickness	6.5	3.4	6.2	1.2 - 14.9	6.0	3.7	5.4	0.1 - 37.5
Platform angle	74.8	6.3	75.0	60.0 - 85.0	72.1	8.6	73	2.0 - 103.0

In summary, knappers of Shurmai Rockshelter tried to produce longer and larger flakes. They did not favor practicing platform adjustment in order to produce this desired end product; instead they increased platform thickness, adjusted platforms and controlled force. My analysis implies that the larger flakes produced early in the core reduction sequence were used first as tools and those smaller flakes, unmodified flakes, produced in late core reduction sequence were stored for the future.

Kakwa Lelash (GnJm 2) Rockshelter. Plain platforms comprise 57.1 percent (N ' 4), retouched platforms 14.3 percent (N ' 1), and miscellaneous 28.6 percent (N ' 2) percent of the flake tool assemblage at Kakwa Lelash. The dominant platform is plain. As 28.6 percent of flake tools are not identified, it is necessary to analyze unmodified flake assemblage in order to understand what type of platforms dominated the flake assemblage of the site.

The unmodified complete flake assemblage from Kakwa Lelash site consists of 3.4 percent (N ' 20) cortical, 50.3 percent (N ' 295) plain, 13.3 percent (N ' 78) faceted, 18.4 percent (N ' 108) retouched, 13.3 percent (N ' 78) broken, and 1.4 percent (N ' 8) miscellaneous platforms. As in the tool assemblage, the dominant platform is the plain type. Faceted platforms, including both faceted and retouched types, are also significant (31.7 percent). Therefore, knappers of both sites preferred flakes with plain platform as their desirable end product (Table 21). However, the preference to plain forms in the Kakwa Lelash is higher than that of the Shurmai. At the same time Kakwa Lelash Rockshelter shows significant decrease in retouched platforms in both tool and unmodified flake assemblages. Retouching the platform is usually considered as useful means of adjusting the exterior platform angle and thereby controlling the distribution of force to remove flakes from a core. However, the use of the retouched platform appears to have decreased through time. This was probably because the stone artifacts of the Later Stone Age in East Africa decreased in size to become microlithics. These smaller stone artifacts were used as a part of composite tools. It appears that the production of longer and larger flake tools by retouching platform decreased.

The size of the complete flake tool and unmodified flake assemblages and platform thickness were measured (Table 22). According to this analysis, flake tools at Kakwa Lelash Rockshelter are generally longer, thicker, andsignificantly narrower than unmodified flakes. Thus, it appears that many flake tools were produced early in the core reduction sequence, while unmodified flakes later in the core reduction sequence. However, the platform thickness of flake tools is smaller than that of unmodified flakes. In this case, knappers of Kakwa Lelash increased exterior platform angles to produce longer flakes. As shown Table 28, the exterior platform angle of flake tools is considerably greater than that of unmodified flakes. As

a result, the platform thickness of the former becomes more variable than that of the latter.

Compared to Shurmai site, the stone artifacts discovered from Kakwa Lelash site are significantly smaller. However, knappers at both sites preferred the longer and larger flakes as tools. In order to produce this desirable end product, the knappers of Shurmai increased the platform thickness, while those of Kakwa Lelash increased exterior platform angles. They also decreased in the adjustment of their platforms through time.

Shape of Tool Blanks. The most important indices for understanding tool blank shape are the value of length bywidth and width by thickness. These indices indicate whether the tool blanks were broad and short or long. According to Table 23, in general, the values of length by width for flake tools are larger than those for unmodified flakes. These indices confirm that flake tools are longer than unmodified flakes. Flake tools recovered from Kakwa Lelash are more elongated than those from Shurmai Rockshelter. However, these indices indicate that the flakes discovered at both sites tend to be short and broad based. Therefore, the initial tool blanks or cobbles would be short and broad in shape and not be long enough to produce long elongated flakes or knappers of the sites would produce short and broad flakes intentionally.

Core Reduction Sequence. The best information on the initial core reduction process is obtained from secondary flakes rather than from primary flakes. As expected, unidirectional flake scar patterns are dominant. Therefore, in the initial reduction sequence, the flakes at Shurmai Rockshelter were generally reduced unidirectionally from single platform core. In practice, this systematic reduction method of unidirectional flake removal from a single platform is a reduction method that maximizes use of raw material and results in the production of longer flakes.

Tertiary flakes and discarded cores can provide useful information about the later stages of the core reduction sequence. Complete tertiary flakes recovered from the Shurmai Rockshelter comprise 60. 1 percent (N ' 507) of unmodified flake and 78.1 percent (N ' 25) of flake tool assemblages. In order to understand whether core reduction and tool blank preparation was changed, flake scar pattern of these tertiary flakes were analyzed. Tertiary flakes consist of 38.9 percent (N ' 197) unidirectional, 24.9 percent (N ' 126) radial, and 3.2 percent (N ' 16) bi-directional flake scar pattern. Unfortunately, 25.6 percent (N ' 130) of flakes could not be identified. As a result, discarded cores were also analyzed. This analysis has already been described in Table 21. The core analysis reveals that radial complete cores comprise 85.2 percent (N ' 46) of the core assemblage (Table 24). The shift from unidirectional to radial flake removal would happen when the core could not produce flakes from a single platform.

G-Young Gang

Table 21. Platform Preparation for Tool Blank of Kakwa Lelash (GnJm 2) Rockshelter

Platform type		Complete flake tools		Unmodified complete flakes	
		Frequency	**Percentage**	**Frequency**	**Percentage**
	Cortical	0	0.0	20	3.4
	Plain	4	57.1	295	50.3
	Faceted	0	0.0	78	13.3
	Retouched	1	14.4	108	18.4
	Broken	0	0.0	78	13.3
	Undetermined	2	28.6	8	1.41
Total		7	100.0	587	100.1

Table 22. Comparison of Flake Tool Size and Unmodified Complete Flake Size of Kakwa Lelash (GnJm 2) Rockshelter

	Complete flake tools				Unmodified complete flakes			
	Mean	(S.d)	Median	Range	Mean	(S.d)	Median	Range
Length	26.0	7.0	28.9	13.7 - 32.4	23.8	7.6	22.7	7.9 - 58.1
Width	13.0	3.5	14.2	6.8 - 16.9	20.5	7.1	19.1	1.5 - 49.5
Thickness	6.7	2.6	7.6	2.8 - 9.2	6.5	2.8	6.0	2.1 - 23.6
Platform thickness	3.6	3.2	2.0	1.3 - 9.0	3.8	1.8	3.5	1.1 - 12.2
Platform angle	82.2	2.17	81.0	80.0 - 85.0	76.8	7.9	77.0	48.0 - 102.0

Table 23. Comparison of Flake Tool Shape and Unmodified Flake Shape

Site		Complete flake tools				Unmodified complete flakes			
Shurmai (GnJm 1) Rockshelter		Mean	(S.d)	Median	Range	Mean	(S.d)	Median	Range
	L/W [1]	1.5	0.9	1.4	0.1 - 5.2	1.3	0.6	1.2	0.3 - 5.8
	W/Th [2]	3.5	1.1	3.4	1.2 - 5.8	3.6	1.2	3.5	0.3 - 19.1
Kakwa Lelash (GnJm 2) Rockshelter	L/W [1]	2.0	0.4	2.0	1.4 - 2.7	1.2	0.5	1.2	0.5 - 8.7
	W/Th [2]	2.2	0.7	2.0	1.5 - 3.5	3.4	1.0	3.3	0.3 - 7.9

[1]: Length by width
[2]: Width by thickness

Kakwa Lelash (GnJm 2) Rockshelter. Secondary flakes constitute only 11.2 percent (N ' 66) of the unmodified complete flake assemblage. Many of these flakes show the remnants of cortex on their distal surfaces (28.8 percent, N ' 19), left lateral (24.2 percent, N ' 16), and on right lateral (22.7 percent, N ' 15) exterior surfaces. As mentioned above, the common characteristic of these cortex remnants indicate that they the flakes were removed from a single platform core. As expected, of these 50 secondary flakes, 82 percent (N ' 41) of flakes were removed by the unidirectional method.

Tertiary flakes and discarded cores were also analyzed. Complete tertiary flakes discovered from Kakwa Lelash Rockshelter make up 86.9 percent (N ' 510) of the unmodified flake and 100.0 (N ' 7) percent of the flake tool assemblages. These tertiary flakes consist of 42.4 percent (N ' 261) unidirectional, 12.5 percent (N ' 64) radial, and 4.3 percent (N ' 22) bi-directional flake scar patterns. However, 27.5 percent (N '140) of flake scar pattern were not identified. As a result, discarded cores were analyzed. Of 25 complete cores, cores with radial scar pattern comprise 76.0 percent (N ' 19) and bi-directional 12 percent (N ' 3) of complete core assemblage. Interestingly, there is not a single unidirectional core in this assemblage. In contrast to the Shurmai Rockshelter, flakes with unidirectional flake scar pattern increased, while those of radial flake scar pattern decreased. However, there are no differences in core assemblage between two sites. Therefore, a similar core reduction sequence is assumed to have been practiced at both sites (Table 25).

Typological Analysis. As noted in the Chapter II, the MSA lithic assemblages in Africa are characterized by flake and blade tools such as points and scrapers and by the absence of the handaxes and cleavers that had dominated the Earlier Stone Age. In order to understand typological changes through time, the assemblages discovered from our two rockshelters are compared. Most typological analyses of flakes, cores and tools have already been described in this techno-morphological analysis section of the present chapter. Therefore, in this section, only the typological analysis of flakes is described.

Shurmai Rockshelter (GnJm 1). Of total analyzed 4,782 stone artifacts, 1467 (30.7 percent) unmodified flakes, 100 (2.1 percent) cores, and 79 (1.7 percent) tools of lithic assemblage discovered from Shurmai Rockshelter (GnJm 1) are analyzed. In terms of flake shape, Sub-triangular and truncated triangular flakes comprise 28.5 percent (N ' 240) of flake assemblage, while convergent flakes contribute 28.2 percent (N ' 238) of assemblage. The assemblage also produced 139 (16.5 percent) blades. Although the assemblage contains flakes of many different shapes, triangular and sub-triangular types dominate (56.7 percent). The assemblage also contains flakes categorized as "backed knife." Of these naturally backed knives, 29 knives are backed in right lateral edge, while 48 knives in left lateral edge. These naturally backed knives comprise

9.1 percent (N ' 77) of the flake assemblage. Figure 16 shows various types of flakes recovered from Shurmai Rockshelter.

Kakwa Lelash (GnJm 2) Rockshelter. The lithic assemblage from the Kakwa Lelash Rockshelter consists of 1,067(13.6 percent) flakes, 32 (0.4 percent) cores, 26 (0.3 percent) tools and 6,737 (85.6 percent) debris. Of 1067 unmodified flakes, sub-triangular flakes comprise 10.1 percent (N ' 59) of total flake assemblage. Convergent flakes contribute 29.5 percent (N ' 173) of the assemblage. Compared with Shurmai Rockshelter, the Kakwa Lelash Rockshelter produced fewer convergent flake forms. In addition, blade and naturally backed knife production is there also low (6.1 percent, N ' 36 and 7.6 percent, N ' 5 respectively). The flakes produced at the Kakwa Lelash Rockshelter are characterized by great variation in type (Figure 17). The flake illustrations were drawn to the same scale. Table 26 summarizes the tool inventories recovered from the two rockshelters.

Functional Analysis

Functional variation between stone tools is commonly reflected in variation in edge angle and edge wear. In order to understand the function of the various tools discovered at the two rockshelters, these two variables were measured. Unfortunately, many tools recovered from Shurmai (GnJm 1) Rockshelter were not shipped into Texas A&M University and therefore had to be studied from line drawings. The edge angle of these tools could be not obtained. The following analysis is based on only those tools that were shipped into the United States.

Edge Angle

Although Shurmai Rockshelter (GnJm 1) site yielded 79 tools, angles of only 13 tools could be measured. The edges of each of these tools were measured at each one-centimeter interval along their margins provided the tool was more than 2 cm long. After these individual measurements were made, the mean edge angles was computed and recorded. When an edge presented a denticulate appearance, each gouge-point was measured as one angle; angles in the concavities were measured as the other angle. After measuring these two angles, the mean of the two angles was computed and recorded as the edge angle for the specimen (Table 27). Edge angles of 6 tools ranged between 50 to 59°. Angles of four tools between 40 to 49°. There was one tool each with edge angles of 20° s, 60° s, and 70° s. Thus, the dominant edge angles of tools recovered from the site ranged between 50 and 59°.
Kakwa Lelash Rockshelter (GnJm 2) produced 26 tools. Of these, the edge angles of 21 tools were measured. The tools with edge angle of 40 to 49° comprise 28.7 percent (N ' 6) of the assemblage, while those with 50 to 59° contribute 23.8 percent (N ' 5). Tools with edge angle of 60 to 69° contribute 19.0 percent (N ' 4) of the total.

Comparing to the tools recovered from the two rockshelters on this dimension reveals only a slight, and no doubt insignificant, difference (Table 28). At Shurmai Rockshelter, the dominant tool edge angle range is 48 to 52°, while Kakwa Lelash it is between 45 to 48° (Figure 18). The Shurmai Rockshelter did not produce tools with edge angle of 30° s, Kakwa Lelash Rockshelter did produce tools with this angle (N ' 3).

Neither site produced any tools with edge angle less than 21°, nor more than 78°. As noted earlier (Gould, Koster, and Sontz 1971:Table 1, Table 2; Hayden 1979a:124-125; Wilmsen 1968:156-158), tools with edge angles less than 20 ° are very rare because those angles are too acute to perform tasks well. In general, acute angle of 26° to 35°

are found on cutting tools, angle of 35° to 40° on for whittling knives, medium angles ranging between 46° and 55° on general purpose tools, and angles of 66° to 75° on heavy duty tools.

Based upon the analysis, the most frequent edge angles of tools discovered from the two sites fall between 45° to 51°. That is, most of tools appear to be general-purpose tools. According to Wilmsen (1967:140), general-purpose tools are those that are used for skinning and hide scraping, sinew and plant fiber shredding, heavy cutting, and tool back blunting. In order to understand the type of task for which tools were used, microwear analysis was undertaken.

Table 24. Comparison of Complete Secondary and Tertiary Flakes of Shurmai (GnJm 1)Rockshelter

Flake type	Placement of cortex			Flake scar pattern *		
Secondary		Frequency	Percentage		Frequency	Percentage
	Proximal	24	8.8	Uni-direction	120	51.9
	Distal	75	27.4	Bidirection	22	9.5
	Left lateral	92	33.6	Radial	31	13.4
	Right lateral	64	23.4	Random	11	4.8
	All around the edge	10	3.6	Undeterm-ined	47	20.3
	Center of the flake	9	3.3	Total	231	99.9
Tertiary				Uni-direction	197	38.9
				Bidirection	16	3.2
				Radial	126	24.9
				Random	38	7.5
				Undeterm-ined	130	25.6
				Total	507	100.1

* The number of flakes are originated from the total of flakes with cortex on distal, left lateral, and right lateral exterior surface.

Table 25. Comparison of Complete Secondary and Tertiary Flakes of Kakwa Lelash (GnJm 2) Rockshelter

Flake type	Placement of cortex			Flake scar pattern *		
Secondary		Frequency	Percentage		Frequency	Percentage
	Proximal	13	19.7	Uni-direction	41	82.0
	Distal	19	28.8	Bidirection	1	2.0
	Left lateral	16	24.2	Radial	5	10.0
	Right lateral	15	22.7	Random	1	2.0
	All around the edge	1		Undetermined	2	4.0
	Center of the flake	2		Total	50	100.0
Tertiary				Uni-direction	216	42.4
				Bidirection	22	4.3
				Radial	64	12.5
				Random	68	13.3
				Undetermined	140	27.5
				Total	510	100.0

* The number of flakes are originated from the total of flakes with cortex on distal, left lateral, and right lateral exterior surface.

Table 26. Typological Analysis

Site	Flakes		Cores		Tools	
Shurmai (GnJm 1)		Frequency (Percentage)		Frequency (Percentage)		Frequency (Percentage)
	Sub-triangular	240 (28.5)	Complete	54 (54.0)	Borer	1
	Triangular/convergent	238 (28.2)	Fragment	46 (46.0)	Hammerstone	3
	Parallel/blade	139 (16.5)	Multiplatform	49 (85.2)	Hafted point	1
	Naturally backed	77 (9.1)	Single platform	3 (5.6)	Small chopper	2
	Plain butt	408 (48.4)			Naturally backed knife	6
	Faceted/retouched butt	343 (40.7)			points	7
	Unidirectional scar pattern	120 (14.2)			Denticulated flakes	13
	Radial scar pattern	31 (3.7)			Notched flake	1
					Blade	5
					Scrapers	46
Kakwa Lelash (GnJm 2)	Sub-triangular	173 (29.5)	Complete	25 (78.1)	Hammerstone	4
	Triangular/convergent	59 (10.1)	Fragment	7 (21.9)	Hafted point	1
	Parallel/blade	36 (6.1)	Multiplatform	0 (0.0)	Small chopper	1
	Naturally backed	5 (0.9)	Single platform	25 (100.0)	points	3
	Plain butt	295 (50.3)			Blade	6
	Faceted/retouched butt	186 (31.7)			Scrapers	2
	Unidirectional scar pattern	208 (35.4)			Micro flake	6
	Radial scar pattern	71 (12.1)				

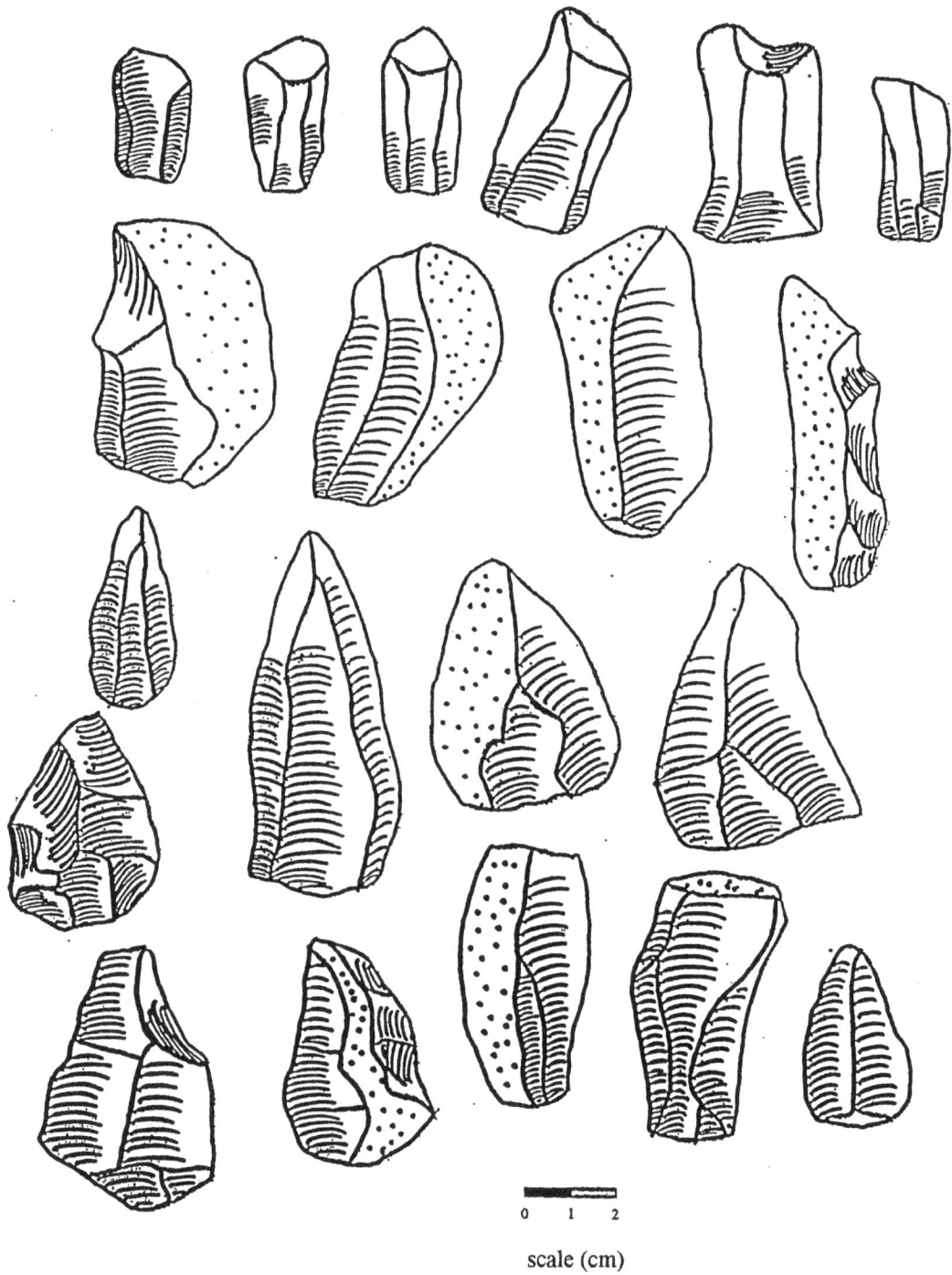

Figure 16. Various Flake Types Recovered from Shurmai Rockshelter (GnJm 1).

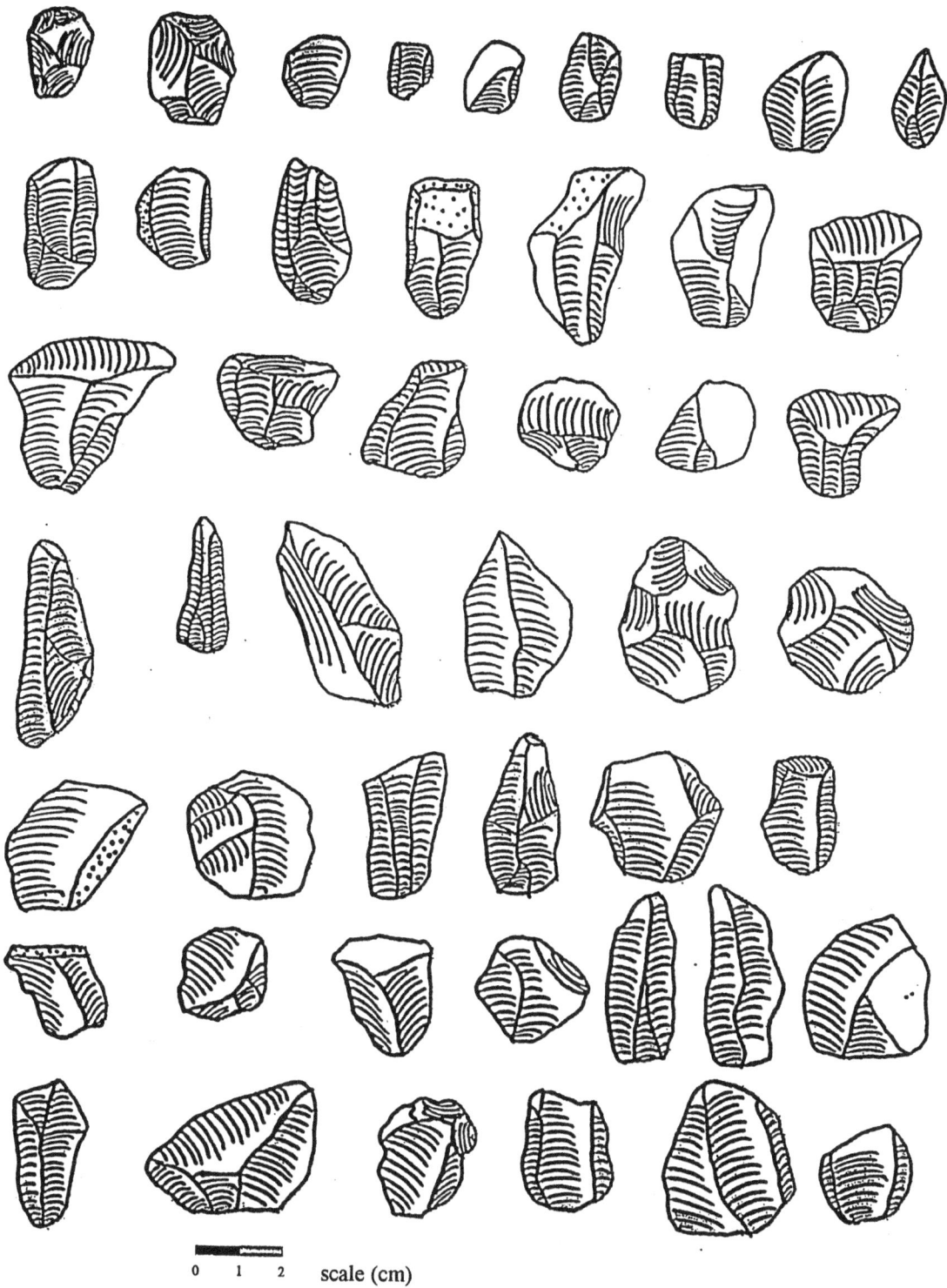

Figure 17. Various Flake Types Recovered from Kakwa Lelash Rockshelter (GnJm 2).

Microwear Analysis

The tools discovered from the two sites were examined by microwear analysis. The type of microscope used for this analysis is described in the Chapter IV. The use-wear analysis in the present study is based on the low power approach. Unfortunately, all tools made of basalt did not reveal any use wear clues to their former function. Although the edges of those tools still show modification, the extreme edges that are the subject of the microwear analysis were severely rounded. This is because those tools were too weathered to have retained any use wear traces. This is unfortunate, as the most common raw material utilized at Shurmai Rockshelter was basalt. Total 79 tools were discovered from Shurmai site. Of these, 39 tools were made of basalt, and 11 tools were made of quartz. In addition, tools not transported to Texas A&M University had to be excluded from this microwear analysis. As a result, only four tools recovered from Shurmai site were found to have analyzable microwear.

Contrast to Shurmai site, Kakwa Lelash Rockshelter produced many tools made of chert and obsidian. In the Kakwa Lelash tool assemblage the extreme edges of tools with edge angle of less than 35° are characterized by small feather scars. In one case, the tool is characterized by polish. In general the tools with this acute angles show scars on lateral extreme edges. Based upon theses characteristic, we may hypothesize that tools with these acute angles would have been used for cutting soft material such as meat, skin, or plant fiber.

Tools with edge angle of more than 40° and less than 60° are characterized by plano-convex shape. The scars on these tools usually appear on their convex or exterior surface. When tools were hafted the scars appear along the distal edge. When tools were not hafted, the scars occurred either one or both lateral and distal edges. In contrast to tools with acute angles, many of these tools of medium angles are carrying step and hinge fractures which are perpendicularly located along the extreme edges or are semicircular shape although some tools are carrying feather scars. Generally these step and hinge fractures are densely located along the tool edge. In some cases, the extreme edges of tools are too blunt or dull to reveal any fractures. These cases suggest that the tools had been either utilized for a long time or in tasks demanding hard pressure. Many of tools with these medium angles were very likely used as scrapers on either hard or soft materials.

Tools with obtuse edge angles of more than 65° reveal some interesting result as well. Microwear analysis of some of the tools with blunt angles shows use wear not on the retouched edge but on the opposite lateral edges characterized by either acute or medium angles. In these cases, the retouched lateral edge was apparently not used as the working edge but as backing for the opposite edge. In sum, it would appear that the Kakwa Lelash tool inventory was dominated by general-purpose tools.

Table 27. Measurement of Edge Angles of Tools

Edge angle (degree)	Shurmai (GnJm 1)		Kakwa Lelash (GnJm 2)	
	Frequency	Percentage	Frequency	Percentage
20 - 29	1	7.7	1	4.8
30 - 39	0	0.0	3	14.3
40 - 49	4	30.8	6	28.7
50 - 59	6	46.1	5	23.8
60 - 69	1	7.7	4	19.0
70 - 79	1	7.7	2	9.5

Table 28. Comparison of Edge Angles

Site	Mean	Standard deviation	Median	Range
Shurmai (GnJm 1)	52.68	11.89	52.70	23.70 - 76.50
Kakwa Lelash (GnJm 2)	52.08	13.14	50.50	29.30 - 76.70

Frequency distribution

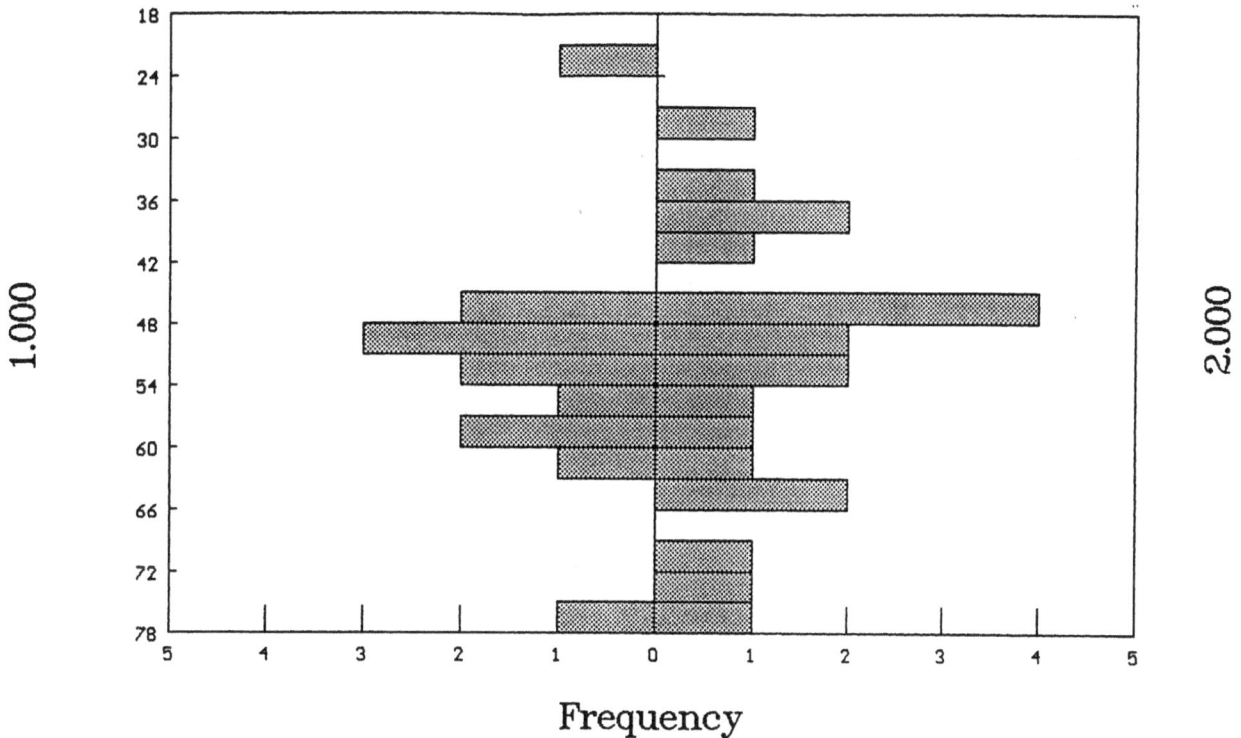

Figure 18. Frequency Distribution of Edge Angles.

Stylistic Analysis

Many scholars suggest that style resides in a distinct and self-contained realm of form exclusive to itself (Binford 1986:560; Close 1978:223). For them, stylistic traits are independent of function. However, Sackett (1973:318) rejects this view. According to him, style does not occupy a distinct domain of form in and of itself. Instead, he asserts that stylistic variation represents a series of specific choices by artisans from a broad spectrum of equally viable alternative ways of achieving the same functional end. According to him function and style are fully complementary. Style and function share equal responsibility for the formal variation exhibited by a given body of artifacts (Sackett 1982:68). In his view, the interpretation of the style cannot and should not be divorced from the consideration of function.

The approach to style used in this work follows Sackett's strictures. Using this approach I examined all modified or retouched tools from the Shurmai and Kakwa Lelash rockshelters stylistically. I found that, although there are various alternative styles in these tools that, at first, appear to be nothing more than expressions of aesthetic attributes, the stylistic variation in these tools was in fact closely related to function in one way or the other.

Shurmai Rockshelter (GnJm 1)

The collection of tools recovered from the Shurmai site consists of one borer, three hammer stones, one hafted point, two small choppers, six naturally backed knives, seven points, 13 denticulated flakes, one notched flake, five blades, and 46 scrapers of various types. Of these tools, only naturally backed knives, points, and scrapers received formal stylistic analysis. As the rest of tool types were represented by only one example, it was not possible to analyze their stylistic variation.

The naturally backed knives collection consists of six examples, four left-laterally backed knives, and two right-laterally backed knives. The two left-laterally backed knives both show modification on their interior surfaces; one shows it on the exterior surface, one on both surfaces. Compared to other tools, the naturally backed knives are relatively large. All six were made on complete flakes that can be categorized as blades. The two right-laterally backed knives show modification on all of their exterior surfaces. Both types of modification indicate that the knives were used in transverse scraping or whittling activities. Both unretouched and retouched backed knives reveal that left-laterally backed knives outnumbered the right-laterally backed ones. The ratio of left to right unretouched backed knives is 48:29, while the ratio is 4:2

53

in the retouched specimens. That is, in both cases, the ratio of left to right laterally backed knives is 2:1. We might be tempted to view this ratio as reflecting purely stylistic, that is, nonfunctional, difference between knife types. However, if this variation is viewed in terms of functional consideration, another explanation is possible. As noted, the backed knives reflect some transverse activities such as scraping, shaving, and whittling. Scraping is motion toward the operator, whittling is motion away from the operator. A right-handed operator would find the right-laterally backed knives easier to use in transverse activities, while left-laterally backed knife would be easier for a left-handed person to use. Therefore, the ratio of left to right specimens in the backed knife collection may be attributable to the variation in ratio of right-handed to left-handed individuals within the MSA population that produced the tools.

The primarily distinction between flake and blade is the ratio of length to width. Particularly long blades are very useful as scrapers. In terms of flake types, the naturally backed knives are secondary flakes. Compared to tertiary flakes, secondary flakes are generally (but not always) longer, larger, and thicker as they are usually produced earlier in the core reduction sequence. That is, the length and completeness of the type of flake on which they are made largely govern the function of the knives. In addition, different types of these flakes result from the operators' initial intentions to get desirable flake tools. Modification also appears only on the lateral edge of the flake. This is because the lateral edge is the longest and thinnest edge that can be most easily utilized as a tool without heavy retouch. Blades produced at the site show very similar style to backed knives although they are generally smaller than those tools. Although such blades generally exhibit use-wear on their lateral edges, unlike backed knives, use-wear also appears on distal part as well. This strongly suggests that these blades served as a part of composite tools. Here again, variation in the types of naturally backed knives and retouched or used blades cannot be divorced from their function.

Seven convergent flake tools or "points" were recovered from Shurmai Rockshelter. Three points are made of chert, two of obsidian, one of quartz, and one of basalt. These "points" do not give any indication of ever having been hafted. However, they do exhibit some variation in modification. While all are characterized by moderate retouch along their edges, the degree of their modification differs depending on the material of their manufacture. On the points made of chert and obsidian, the whole surface (often both surfaces) is carefully retouched. Basalt points are only partially modified on their exterior surfaces. Obsidian points carry only parallel retouch; no modification appears on the proximal part of the obsidian specimens. In the case of "points", stylistic variation clearly relates to the raw material choice made in their manufacture.

The Shurmai Rockshelter also produced 46 side scrapers and end scrapers. The style of these scrapers varies widely. Analysis of these scrapers reveals that their makers utilized the longest natural axis of their flakes. If the flake shape is divergent or square, the tool was used as end scraper. If the flake was long or oval in shape, it was used as a side scraper. Thus, it would appear that the initial size and shape of flake determined the choice of the "style" of the scraper. Figure 19 illustrates various tool types recovered from Shurmai Rockshelter (GnJm 1).

Kakwa Lelash Rockshelter (GnJm 2)

The tool assemblage recovered from Kakwa Lelash consists of four hammer stones, one hafted point, three unhafted points, one very small chopper, two scrapers, points, six blades, and six micro-flakes that exhibit use-wear. Only the points, blades, and micro flake tools were present in sufficient quantities to allow stylistic comparisons.

Of the three unhafted points, two of them are characterized by thick right-lateral edges that render them nearly prismatic in shape. At first I thought their modified right lateral edges showed signs of use wear. However, microwear analysis reveals the opposite: it was the thin left lateral edges that actually exhibit the use wear. The modified right-lateral edges apparently served only as backing. The modification occurs at different part on these Kakwa Lelash points than on the points from the Shurmai Rockshelter. Since the Kakwa Lelash Rockshelter did not produce any naturally backed knife, these points may have been used to perform the same functions as backed knives. The third unhafted point also carries modification. However, this modification appears only on its extreme lateral edge. Here again, this contrasts with the points recovered from Shurmai Rockshelter, which exhibit modification over all the edges except for the proximal part of a flake. All in all these data suggest that, through time, naturally backed knives were replaced by "Points" artificially backed by formal modification. The same function is performed by artifacts of radically different styles.

The blade tools recovered from the Kakwa Lelash Rockshelter are generally rectangular in shape and significantly smaller than those from Shurmai Rockshelter. The modification of these blade flakes usually occurs on the lateral edge. Sometimes, the distal part of the blade flakes is also modified. In other cases, both proximal and distal parts of the blade flakes were removed so that only medial section of a flake was used as a tool. This variation probably reflects the development of microliths for use in composite tools. The development of microlithic technology is one of the significant stylistic distinctions between the Middle and Later Stone ages. A significant portion of Kakwa Lelash tool assemblage consists of such microliths. Their forms include crescentric, oval or circularly shaped flakes. The style of these microliths is less standardized when compared to the flakes of the earlier period. Modification of these various flakes appear only on their lateral or distal part. That is, the position of modification on these flakes is determined by the position of their pointed edges. Therefore, the style of these microliths largely depended on functional consideration. Figure 20 illustrates various types of tools recovered from Kakwa Lelash Rockshelter.

The stylistic analysis of tools recovered from the two archaeological sites demonstrates that, as Sackett suggests, the style of formal tools and their function are closely related. Further, it indicates, in addition to function, that the formal or stylistic variation in these tools is also determined by the raw material and morphology of the flakes themselves.

Chapter V
Conclusion

Two archaeological sites, Shurmai (GnJm 1) and Kakwa Lelash (GnJm 2) rockshelters, Kenya were excavated between 1993 and 1995. The former site was found to contain the Middle to Later Stone Age deposits and the latter consists of the Later Stone Age deposits. In this work, I reported my analysis and comparison of the stone artifacts recovered from the two sites in terms of their raw material, techno-morphological attributes, function, and styles. This work had three goals: (1) to describe the attributes of debitage, tools, and cores from the two sites, (2) to measure and compare the technological change evident at these two rockshelters through time, and (3) to monitor and compare changes in raw material usage at each through time. In these ways I hoped to increase our understanding of the lithic technology of the Middle and Later Stone Age in East Africa.

Both Shurmai (GnJm 1) and Kakwa Lelash (GnJm 2) rockshelters are located near the top of large Pre-Cambrian metamorphic erosional inselbergs made of granite gneiss. Geologically the sites are within an area rich in quartz sources and abundant volcanic rocks including basalt, granite, gneiss, and schist. However, in terms of raw materials for stone tool making, all of these volcanic rocks are as poor quality. Because the degree of modification of stone artifacts is closely related to the quality of raw material, I expected that the inhabitants of the sites would have procured higher quality of raw material from sources far away from the two sites.

Here, the term "high quality" raw materials mean stones that fracture conchoidally and that are homogenous, isotropic, brittle, and elastic. Obsidian and chert are generally considered as high quality materials for knapping. In fact, there is no high quality raw material source areas within 10 km from the two archaeological sites. In order to understand raw material procurement strategies of the Middle and Later Stone ages, the types of raw materials produced from the two sites were analyzed. The analysis reveals that the Middle Stone and Later Stone ages practiced different raw material procurement strategies. The MSA people heavily procured local materials regardless the quality of the stones, while the LSA people used both local and non-local material almost equally. However, the LSA people more considered the quality of raw materials and appear to have preferred high quality raw materials for tool manufacture. Transportation of procured raw materials also shows differences between the two periods. As most of local materials utilized by the MSA people were available within 10 km from the site, the procured raw materials apparently were transported to the site as nodules without being roughed out somewhere else. In contrast, when the LSA people procured raw materials far away from the site, they roughed out blanks before transporting the materials to the site. In addition, the amount of non-local materials produced from the two archaeological sites indicates that the MSA people did not procured non-local material systematically. The MSA

people probably found high quality raw materials only accidentally during hunting, collecting, or seasonal movements. The LSA people, on the other hand, appear to have procured the high quality raw materials in a systematic manner since the amount of high quality non-local materials utilized to produce desirable tools is higher than that of poor quality local materials in the LSA components. In Binford's (1979:259-261) terms, the raw material procurement strategy of the MSA was "embedded," while that of the LSA was "curated." However, the change in raw material procurement strategies from embedded to curated in our study area appears to have been gradual rather than abrupt.

A second contrast between lithic technologies at the MSA and the LSA sites is the evidence of a change in the kinds of flakes preferred in tool manufacture. Knappers of the MSA sites tried to produce larger and longer flakes. In order to get these desirable flakes, the knappers chose to increase platform thickness rather than modify platforms. In general, the larger flakes produced in early core reduction sequence appear to have been preferred as tools although some flakes produced in the later core reduction sequence were no doubt stored for future use.

Unlike the stone artifacts recovered from the MSA site, those from the LSA site show significant decrease in retouched striking platforms in both flake tools and unmodified flakes. In addition, the typo-morphological analysis of the two archaeological sites reveals that the size of artifacts significantly decreased from the MSA to the LSA. As the size of artifacts decreased, core size became smaller and the flake shape became more varied. As a consequence, the cores recovered from the LSA site are more apt to be exhausted than those from the MSA site. However, this exhaustion may not be due entirely to reduced flake size. The relative scarcity and value of cores of non-local material may have caused knappers to try to "use them up" before discarding them. Finally, the MSA site produced many blades and naturally backed knives while the LSA site yielded few blades and no naturally backed knives. This change probably indicates a trend in the LSA in which microliths used in composite tools gradually replaced naturally backed knives in the tool inventory.

In contrast to the foregoing, the functional analysis and comparison of the MSA and the LSA lithic assemblages from the two sites did not reveal significant differences. In general, tools with edge angle ranging 46 to 55 degrees are considered to have served as general-purpose tools. Flakes with edge angle less than 20 degrees are generally too sharp to serve as tools. The majority of the tools produced from the MSA assemblage have edge angles of 50 to 59 degrees while tools recovered from the LSA site exhibit 40 to 49 degrees angles. No tools with edge angle less than 20 degrees were recovered at either site.

Stylistic analysis of the stone artifacts recovered from the two sites was also undertaken. Although neither of the sites produced heavy-duty tools such as handaxe and cleavers, the assemblages from the sites showed other stylistic

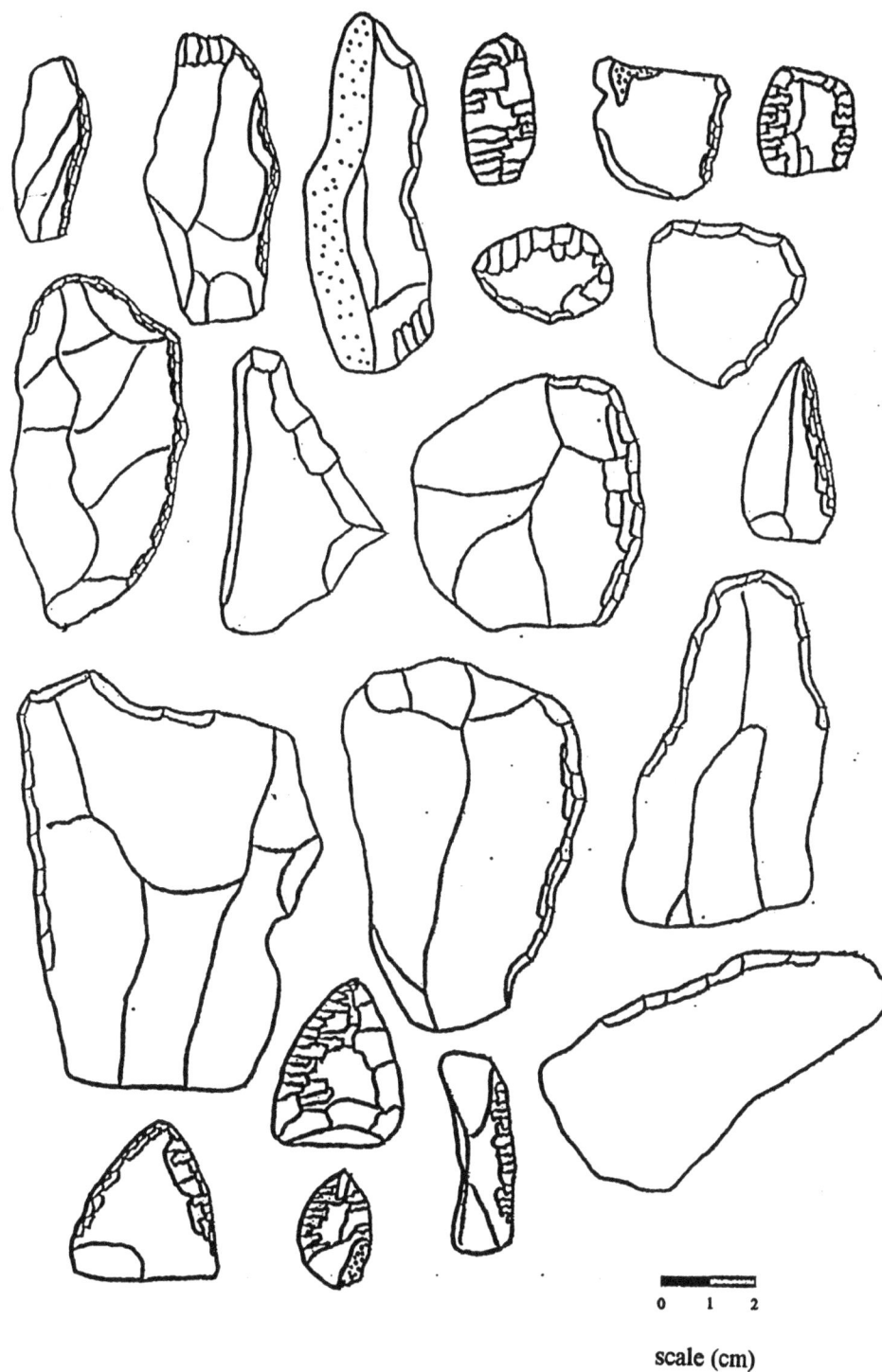

Figure 19. Various Tool Types Recovered from Shurmai Rockshelter (GnJm 1).

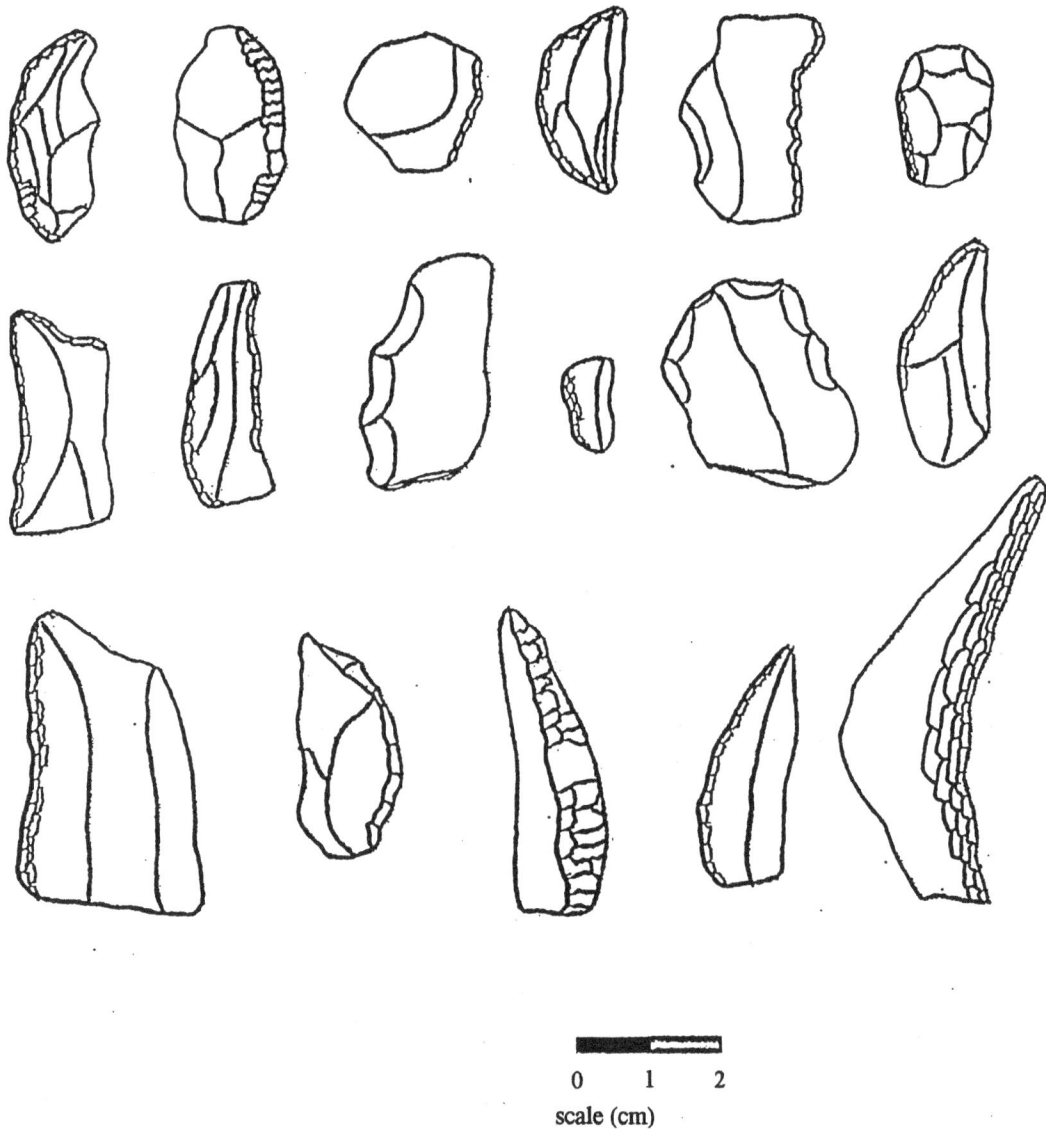

0   1   2

scale (cm)

Figure 20. Various Tool Types Recovered from Kakwa Lelash Rockshelter (GnJm 2).

differences. The MSA assemblage from the Shurmai Rockshelter yielded flake tools such as blades, flake points, flakes with denticulated, and naturally backed knives while that of the LSA at the Kakwa Lelash Rockshelter largely consisted of microlithic blades and flakes with use wear. The production of naturally backed knives and convergent flakes was also low in the LSA components at both sites.

Compared to flakes recovered from the Kakwa Lelash Rockshelter, those recovered from the MSA site are characteristically large and broad. Blades are either irregular or rectangular in shape. The majority of flakes carry plain striking platform instead of the faceted ones more generally characteristic of the MSA in Africa. The lithic industry of the Shurmai site shows attributes similar to those found far away in the Oakhurst Complex industry in South Africa. The Oakhurst Complex appeared in the

later Middle Stone Age and the earlier Later Stone Age there. The MSA component in the present study belongs to the later Middle Stone Age. It is uncertain as to whether the parallels between the MSA lithic industry from Shurmai Rockshelter and the Oakhurst Complex from South Africa are due to the shared characteristics of raw material, to outright diffusion, or to the influence of some other external factors. Future stylistic comparisons of the later MSA lithic industries of East Africa with those of the southern Africa may resolve this question.

In sum, the analyses of raw material and the techno-morphological, functional, and stylistic attributes of the stone artifacts recovered from the Shurmai (GnJm 1) and Kakwa Lelash (GnJm 2) rockshelters revealed both differences and similarities in the raw material procurement strategies, morphological attributes of artifacts, and stylistic variations between the MSA and

58

LSA occupations at these two archaeological sites. However, this analysis found little difference in technology and the functional attributes of the tools at the two sites although, through time, techniques of manufacture become more refined. As Shurmai Rockshelter (GnJm 1) consists of MSA, transitional, and LSA deposits, the site provides us with very useful information about poorly known period in East Africa prehistory. Finally the site strongly suggests that the development of the LSA from the MSA and the changes in these lithic industries were gradual.

Future Research

Numerous questions still remain unanswered by the present study. One of these questions is the nature of subsistence from the MSA to the LSA at the two rockshelters. In addition, a detailed paleoenvironmental study remains to be undertaken. When this information is obtained, a holistic reconstruction of the adaptation of the MSA, the LSA, and the transition from the MSA to LSA will be possible.

Based upon the present analysis, lithic inventories from surface site located sites during 1996 site reconnaissance in the study are yet to be analyzed. When the lithic industries from our two sites are compared with those from these surface sites, our understanding of the MSA and LSA in the study area and in East Africa will be further broadened. In addition, further tests of these conclusions against lithic data to be recovered in future expanded excavations at both caves will also be necessary.

In terms of lithic raw materials, some Mukogodo Masai suggested that exotic material, obsidian in the present study, was found by the site (Cronk 1997, personal communication). Until now we had not found any exotic material source area either at the site or near the site. Future research on the lithic raw material source area is also required.

References Cited

Ackerly, Neal W.
1979 The Southern Desert Study Area. In *An Archaeological Survey of the Cholla-Saguaro Transmission Line Corridor*, assembled by Lynn S. Teague and Linda L. Mayro, pp. 267-406. Arizona State Museum Archaeological Series No. 135. Tucson.

Ahler, S. A.
1989 Mass Analysis of Flaking Debris: Studying the Forest Rather than the Tree. In *Alternative Approaches to Lithic Analysis*, edited by D.O. Henry and G.H. Odell, pp.85-118. Archaeological Papers of the American Anthropological Association No. 1. Washington, DC.

Anderson, Patricia C.
1980 A Testimony of Prehistoric Tasks: Diagnostic Residues on Stone Tool Working Edges. *World Archaeology* 12(2):181-194.

Avery, Graham, Kathryn Cruz-Ubibe, Paul Goldberg, Frederick E. Grine, Richard G. Klein, Michael J. Lenardi, Curtis W. Marean, W. Jack Rink, Henry P. Schwarcz, Anne I. Thackeray, and Michael L. Wilson
1997 The 1992-1993 Excavations at the Die Kelders Middle and Later Stone Age Cave Site, South Africa. *Journal of Field Archaeology* 24:263-291.

Bada, Jeffrey L., and Lydia Deems
1975 Accuracy of Dates beyond the ^{14}C Dating Limit Using the Aspartic Acid Racemization Reaction. *Nature* 255:218-219.

Bates, Robert L., and Julia A. Jackson (editors)
1984 *Dictionary of Geological Terms*. Doubleday, New York.

Baumler, Mark Frederick
1987 *Core Reduction Sequences: An Analysis of Blank Production in the Middle Paleolithic of Northern Bosnia, Yugoslavia*. Ph.D. dissertation. The University of Arizona, Tucson.

Beaumont, P.B., H. De Villiers, and J.C. Vogel
1978 Modern Man in Sub-Saharan Africa prior to 39,000 Years B.P.: A Review and Evaluation with Particular Reference to Border Cave. *South African Journal of Science* 74:409-419.

Bennett, C.F.
1979 Radiocarbon Dating with Accelerators. *American Scientist* 67:450-457.

Berger, Michael Edward
1971 *Habits and Population Parameters of Olive Baboons in the Laikipia District of Kenya*. M.S. Thesis. Texas A&M University, College Station.

Binford, Lewis R.
1963 Red Ocher Caches from the Michigan Area: A Possible Case of Cultural Drift. *Southwestern Journal of Anthropology* 19:89-108.

1979 Organization and Formation Processes: Looking at Curated Technologies. *Journal of Anthropological Research* 35:255-273.

1986 An Alyawara Day: Making Men's Knives and Beyond. *American Antiquity* 51(3):547-562.

Bishop, W.W., and J.D. Clark (editors)
1967 *Background to Evolution in Africa*. University of Chicago Press, Chicago.

Boëda, Eric
1995 Levallois: A Volumetric Construction, Methods, and Technique. In *The Definition and Interpretation of Levallois Technology*, edited by H.L. Dibble and O. Bar-Yosef, pp. 41-68. Prehistory Press, Madison, WI.

Bordes, Francois
1961 Mousterian Cultures in France. *Science* 134:803-810.

1968 *The Old Stone Age*. McGraw-Hill Book Company, New York.

Bradley, Bruce A.
1975 Lithic Reduction Sequences: A Glossary and Discussion. In *Lithic Technology: Making and Using Stone Tools*, edited by E. Swanson, pp. 5-14. Mouton, Chicago.

Brooks, A., D. Helgren, J. Cramer, A. Franklin, W. Hornyak, J. Keating, R. Klein, W. Rink, H. Schwarcz, J. Smith, K. Stewart, N. Todd, J. Verniers, and J. Yellen
1995 Dating and Context of Three Middle Stone Age Sites with Bone Points in the Upper Semliki Valley, Zaire. *Science* 268:548-553.

Brooks, A.S., P.E. Hare, J.E. Korkis, G.H. Miller, R.D. Ernst, and F. Wendorf
1990 Dating Pleistocene Archaeological Sites by Protein Diagenesis in Ostrich Eggshell. *Science* 248:60-64.

Brown, Gary
1991 Embedded and Direct Lithic Resource Procurement Strategies on Anderson Mesa. *Kiva* 56(4):359-384.

Brown, L.H.
1963   *A National Cash Crops Policy for Kenya.* Government Printer, Nairobi.

Busch, Richard (editor)
1993   *Laboratory Manual in Physical Geology.* Macmillan Publishing Company, New York.

Butzer, K.W.
1979   Comments on Implications of Border Cave Skeletal Remains for Later Pleistocene Human Evolution. *Current Anthropology* 20(1):28.

1982   Geomorphology and Sediment Stratigraphy. In *The Middle Stone Age at Klasies River Mouth in South Africa*, edited by R. Singer and J. Wymer, pp.33-43. The University of Chicago Press, Chicago.

1984   Archaeogeology and Quaternary Environment in the Interior of Southern Africa. In *Southern African Prehistory and Paleoenvironments*, edited by R.G. Klein, pp.1-64. A.A. Balkema, Rotterdam.

1987   *Comments on Dating the Hominid Remains from Border Cave and Omo Kibish.* Paper presented in the Present Symposium.

Butzer, K.W., P.B. Beaumont, and J.C. Vogel.
1978   Lithostratigraphy of Border Cave, KwaZulu, South Africa: A Middle Stone Age Sequence Beginning c. 195,000 B.P. *Journal of Archaeological Science* 5:317-341.

Callahan, Errett
1994   *Primitive Technology: Practical Guidelines for Making Stone Tools, Pottery, Basketry, Etc. the Aboriginal Way.* Piltdown Productions, Lynchburg, VA.

1996   *The Basics of Biface Knapping in the Eastern Fluted Point Tradition: A Manual for Flintknappers and Lithic Analysts.* Piltdown Productions, Lynchburg, VA.

Cann, Rebecca L., Mark Stoneking, and Allan C. Wilson
1987   Mitochondrial DNA and Human Evolution. *Nature* 325:31-36.

Chapman, Richard C.
1977   Analysis of the Lithic Assemblages. In *Settlement and Subsistence along the Lower Chaco River*, edited by Charles A. Reher, pp. 371-452. The University of New Mexico Press, Albuquerque.

Charsley, T.J.
1989   Composition and Age of Older Outwash Deposits along the Northwestern Flank of

Mount Kenya. In *Quaternary and Environmental Research on East African Mountains*, edited by W.C. Mahaney, pp. 165-174. York University, Toronto.

Clark, J. Desmond
1954   *Prehistoric Cultures of the Horn of Africa.* Cambridge University Press, Cambridge.

1969   *Kalambo Falls Prehistoric Site.* Cambridge University Press, Cambridge.

1970   *The Prehistory of Africa.* Thames and Hudson, New York.

Clark, J. Desmond., and C. Vance Haynes
1970   An Elephant Butchery Site at Mwanganda's Village Karonga, Malawi. *World Archaeology* 1:390-411.

Clark, J. Desmond, and Kenneth D. Williamson
1984   A Middle Stone Age Occupation Site at Porc Epic Cave, Dire Dawa Part I. *The African Archaeological Review* 2:37-64.

Close, Angela E.
1978   The Identification of Style in Lithic Artifacts. *World Archaeology* 10:223-237.

Coetzee, J.A., and E.M. van Zinderen Bakker
1989   Paleoclimatology of East Africa during the Last Glacial Maximum: A Review of Changing Theories. In *Quaternary and Environmental Research on East African Mountains*, edited by W.C. Mahaney, pp. 189-198. York University, Toronto.

Cole, Sonia
1963   *The Prehistory of East Africa.* MacMillan, New York.

Collins, Michael B.
1975   Lithic Technology as a Means of Processual Inference. In *Lithic Technology: Making and Using Stone Tools*, edited by E. Swanson, pp. 15-34. Mouton, Chicago.

1993   Comprehensive Lithic Studies: Context, Technology, Style, Attrition, Breakage, Use-Wear and Organic Residues. *Lithic Technology* 18 (1 and 2):87-94.

Cooke, C.K.
1963   Report on Excavations at Pomongwe and Tshangula Caves, Matopo Hills, Southern Rhodesia. *South African Archaeological Bulletin* 18:73-151.

1969   A Reexamination of the Middle Stone Age Industries of Rhodesia, *Arnoldia* 6 (20).

Cooke, H.B.S., B.D. Malan, and L.H. Wells
1945 Fossil Man in the Lebombo Mountains, South Africa: The Border Cave, Ingwavuma District, Zululand. *Man* 45:6-13.

Copeland, L.
1983 Levallois/Non-Levallois Determinations in the Early Levant Mousterian: Problems and Questions for 1983. *Paléorient* 9:15-27.

Cotterell, Brian, and John Kamminga
1979 The Mechanics of Flaking. In *Lithic Use-Wear Analysis*, edited by Brian Hayden, pp. 97-112. Academic Press, New York.

1987 The Formation of Flakes. *American Antiquity* 52(4):675-708.

Crabtree, Don E.
1970 Flaking Stone with Wooden Implements. *Science* 169:146-153.

1972 *An Introduction to Flintworking.* Idaho State University Museum Occasional Papers No. 28. Pocatello.

1975 Comments on Lithic Technology and Experimental Archaeology. In *Lithic Technology: Making and Using Stone Tools*, edited by E. Swanson, pp. 105-114. Mouton, Chicago.

Cronk, Lee
1989 *The Behavioral Ecology of Change among the Mukogodo of Kenya.* Ph.D. Dissertation. Northwestern University, Evanston, IL.

Day, M.H., M.D. Leakey, and C. Magori
1980 A New Hominid Fossil Skull (L.H. 18) from the Ngaloba Beds, Laetoli, Northern Tanzania. *Nature* 284:55-56.

Deacon, H.J.
1975 Demography, Subsistence and Culture during the Acheulian in Southern Africa. In *After the Australopithecines*, edited by K.W. Butzer and G. Ll. Isaac, pp. 543-569. Mouton, Chicago.

1979 Excavations at Boomplaas Cave: A Sequence through the Upper Pleistocene and Holocene in South Africa. *World Archaeology* 10(3):241-257.

Deacon, J.
1984 Later Stone Age People and Their Descendants in Southern Africa. In *Southern African Prehistory and Paleoenvironments*, edited by R.G. Klein, pp. 221-328. A.A. Balkema, Rotterdam.

Debénath, André, and Harold L. Dibble
1994 *Handbook of Paleolithic Typology.* University of Pennsylvania, Philadelphia.

deMenocal, Peter
1995 Plio-Pleistocene African Climate. *Science* 270(6):53-59.

Dibble, Harold L., and John C. Whittaker
1981 New Experimental Evidence on the Relation between Percussion Flaking and Flake Variation. *Journal of Archaeological Science* 8:283-298.

Dickson, D. Bruce
1993a *An Ethnoarchaeological Site Survey of Dry Rockshelters Formerly Occupied by the Mukogodo Maasai Hunter-Gatherers of Eastern Laikipia District, Kenya, East Africa.* 58th Annual Meeting of the Society for American Archaeology, St. Louis, MO.

1993b *Ancient Preludes: World Prehistory from the Perspectives of Archaeology, Geology, and Paleoecology.* West Publishing Company, St. Paul, MN.

1995 *Late Quaternary Human Occupation, Paleoenvironment and Landscape Development in the Eastern Laikipia District of Kenya, East Africa.* Proposal Submitted to the National Science Foundation.

Dickson, D. Bruce, and David D. Kuehn
1995 *Prehistory of the Mukogodo Hills: Late Quaternary Human Occupation of the Shurmai Rockshelter (GnJm 1), Eastern Laikipia District, Kenya.* 94th Annual Meeting of the American Anthropological Association, Washington, DC.

1997 *A Preliminary Outline of Late Quaternary Paleoenvironment and Human Occupation of the Mukogodo Hills of Southern Isiolo and Northeastern Laikipia District, Kenya.* 62nd Annual Meeting of the Society for American Archaeology, Nashville, TN.

Dickson, D. Bruce, David D. Kuehn, Lee Cronk
1995 *Ethnoarchaeology of the Mukogodo of Laikipia District, Kenya. Report to the Office of the President of the Republic of Kenya.* Department of Anthropology and the Center for Environmental Archaeology, Texas A&M University, College Station.

Ehret, Christopher
1974 *Ethiopians and East Africans: The Problem of Contacts.* East African Publishing House,

Nairobi.

1976   Cushitic Prehistory. In *The Non-Semitic languages of Ethiopia*, edited by M.L. Bender, pp. 85-96. Monograph No. 5, Occasional Papers Series. East Lansing: Committee on Ethiopian Studies, African Studies Center, Michigan State University. Lansing.

Fedje, Daryl
1979   Scanning Electron Microscopy Analysis of Use-Striae. In *Lithic Use-Wear Analysis*, edited by B. Hayden, pp. 179-188. Academic Press, New York.

Fish, Paul R.
1979   *The Interpretive Potential of Mousterian Debitage.* Anthropological Research Paper no. 15. Arizona State University, Tempe.

1981   Beyond Tools: Middle Paleolithic Debitage Analysis and Cultural Inference. *Journal of Anthropological Research* 37:374-386.

Flury, Manuel
1988   Small-Scale Farming and Changes of Land Use in the Highland of Laikipia, Kenya. *Mountain Research and Development* 8(4):265-272.

Fock, G.J.
1968   Rooidam, a Sealed Site of the First Intermediate. *South African Journal of Science* 64:153-159.

Goodwin, A.J.H.
1926   South African Stone Implement Industries. *South African Journal of Science* 23:784-788.

1946   Earlier, Middle, and Later. *South African Archaeological Bulletin* 1:74-76.

Goodwin, A.J.H., and C. van Riet Lowe
1929   The Stone Age Cultures of South Africa. *Annals of the South African Museum* 27:1-289.

Gould, Richard A., D.A. Koster, and A.H. Sontz
1971   The Lithic Assemblage of the Western Desert Aborigines of Australia. *American Antiquity* 36:149-169.

Gramly, Richard Michael
1976   Upper Pleistocene Archaeological Occurrences at Site GvJm/22, Lukenya Hill, Kenya. *Man* 11:319-344.

Greenberg, J. H.
1963   The Mogododo, a Forgotten Cushitic People. *Journal of African Languages* 2:29-43.

Gregory, J. W.
1896   *The Great Rift Valley: Being the Narrative of a Journey to Mount Kenya and Lake Baringo.* Frank Cass and Co. Ltd., London.

Grün, Rainer, Nicholas J. Shackleton, and Hilary J. Deacon
1990   Electron-Spin-Resonance Dating of Tooth Enamel from Klasies River Mouth Cave. *Current Anthropology* 31(4):427-432.

Hamilton, A. C.
1982   *Environmental History of East Africa: A Study of the Quaternary.* Academic Press, New York.

Hayden, B.
1979a   *Paleolithic Reflections: Lithic Technology and Ethnographic Excavations among Australian Aborigines.* Humanities Press, New York.

1979b   (editor) *Lithic Use-Wear Analysis.* Academic Press, New York.

1993   The Cultural Capacities of the Neanderthals: A Review and Reevaluation. *Journal of Human Evolution* 24:113-146.

Hayden, B., N. Franco, and J. Spafford
1996   Evaluating Lithic Strategies and Design Criteria. In *Stone Tools: Theoretical Insights into Human Prehistory*, edited by George H. Odell, pp. 9-50. Plenum Press, New York.

Hendey, Q.B., and T.P. Volman
1986   Last Interglacial Sea Levels and Coastal Caves in the Cape Province, South Africa. *Quaternary Research* 25:189-198.

Huckell, Bruce B.
1973   The Gold Gulch Site: A Specialized Cochise Site near Bowie, Arizona. *The Kiva* 39:105-129.

Inizan, Marie-Louise, Hélène Roche, and Jacques Tixier
1992   *Technology of Knapped Stone.* Cercle de Recherches et d'Etudes Prehistoriques, Meudon, France.

Jelinek, Arthur J.
1976   Form, Function, and Style in Lithic Analysis. In *Cultural Change and Continuity: Essays in Honor of James Bennett Griffin*, edited by Charles E. Cleland, pp. 19-33. Academic Press, New York.

Jones, Steve, Robert Martin, and David Pilbeam (editors)
1992   *The Cambridge Encyclopedia of Human Evolution.* Cambridge University Press. Cambridge.

Keeley, L.H.
1980   *Experimental Determination of Stone Tools: A Microwear Analysis.* University of Chicago Press, Chicago.

Keeley, L.H., and M.H. Newcomer
1977   Microwear Analysis of Experimental Flint Tools: A Test Case. *Journal of Archaeological Science* 4:798-809.

Keller, C.M.
1969   Mossel Bay: A Redescription. *South African Archaeological Bulletin* 23:131-140.

1973   *Montagu Cave in Prehistory: A Descriptive Analysis.* Anthropological Records, Vol. 28. University of California Press, Berkeley.

Klein, Richard
1975   Middle Stone Age Man-Animal Relationships in Southern Africa: Evidence from Die Kelders and Klasies River Mouth. *Science* 190:265-267.

1989   *The Human Career: Human Biological and Cultural Origins.* University of Chicago Press, Chicago.

1995   Anatomy, Behavior, and Modern Human Origins. *Journal of World Prehistory* 9(2):167-198.

Ku, T.-L, M.A. Kimmel, H.H. Easton, and T.J. O'Neil
1974   Eustatic Sea Level 120,000 Years Ago on Oahu, Hawaii. *Science* 183:959-962.

Kuehn, D.D., and D. Bruce Dickson
1997   *Falling Rocks and Flowing Water: Stratigraphy and Archaeological Site Formation at the Shurmai Rockshelter (GnJm 1), North Central Kenya.* 62nd Annual Meeting of the Society for American Archaeology, Nashville, TN.

Kuman, Kathleen, and R.J. Clarke
1986   Florisbad: New Investigations at a Middle Stone Age Hominid Site in South Africa. *Geoarchaeology* 1(2):103-125.

Lawn B.R., and D.B. Marshall
1979   Mechanisms of Microcontact Fracture in Brittle Solids. In *Lithic Use-Wear Analysis*, edited by B. Hayden, pp. 63-82. Academic Press, New York.

Lawrence, Robert A.
1979   Experimental Evidence for the Significance of Attributes Used in Edge-Damage Analysis. In *Lithic Use-Wear Analysis*, edited by B.

Hayden, pp. 113-122. Academic Press, New York.

Leakey, L.S.B.
1931   *The Stone Age Cultures of Kenya Colony.* Cambridge University Press, London.

Leakey, R., K.W. Butzer, and Michael H. Day
1969   Early *Homo sapiens* Remains from the Omo River Region of Southwest Ethiopia. *Nature* 222:1132-1138.

Lewin, R.
1981   Ethiopian Stone Tools Are World's Oldest. *Science* 211:806-807.

Luedtke, B.E.
1992   *An Archaeologist's Guide to Chert and Flint.* Archaeological Research Tool 7. Institute of Archaeology, University of California, Los Angeles.

Magori, G.C., and M.H. Day
1983   Laetoli Hominid 18: An Early *Homo sapiens* Skull. *Journal of Human Evolution* 12:747-753.

Malan, B.D., and H.B.S. Cooke
1941   A Preliminary Account of the Wonderwerk Cave, Kuruman. *South African Journal of Science* 37:300-321.

Mandeville, M.D., and J.J. Flenniken
1974   A Comparison of the Flaking Qualities of Nehawka Chert before and after Thermal Pretreatment. *Plains Anthropologist* 19(64):146-148.

Mason, R.J.
1962   *The Prehistory of the Transvaal.* Witwatersrand University Press. Johannesburg.

Masse, W. Bruce
1980   *Excavations at Gu Achi: A Reappraisal of Hohokam Settlement and Subsistence in the Arizona Papagueria.* Western Archaeological Center Publications in Anthropology 12. Tucson, AZ.

McBrearty, Sally
1988   The Sangoan-Lupemban and Middle Stone Age Sequence at the Muguruk Site, Western Kenya. *World Archaeology* 19(3):388-420.

1990   Consider the Humble Termite: Termites as Agents of Post-depositional Disturbance at African Archaeological Sites. *Journal of Archaeological Science* 17:111-143.

Mehlman, Michael J.
1977 Excavations at Nasera Rock, Tanzania. *Azania* 12:111-118.

1987 Provenience, Age and Association of Archaic *Homo Sapiens* Crania from Lake Eyasi, Tanzania. *Journal of Archaeological Science* 14:133-162.

Mellars, P.
1989 Major Issues in the Emergence of Modern Humans. *Current Anthropology* 30:349-285.

Merrick, H.V., F.H. Brown, and M. Connelly
1990 Sources of the Obsidian at Ngamuriak and Other South-Western Kenyan Sites. In *Early Pastoralists of South-Western Kenya*, edited by P. Robertshaw. British Institute in Eastern Africa Memoir 11:173-182.

Merrick, Harry V., and Francis H. Brown
1984 Obsidian Sources and Patterns of Source Utilization in Kenya and Northern Tanzania: Some Initial Findings. *The African Archaeological Review* 2:129-152.

Michels, Joseph W., Curtis A. Marean
1984 A Middle Stone Age Occupation Site at Porc Epic Cave, Dire Dawa Part II. *The African Archaeological* Review 2:64-71.

Michels, Joseph W., Ignatius S.T. Tsong, and Charles M. Nelson
1983 Obsidian Dating and East African Archaeology. *Science* 219:361-366.

Munday, F.C.
1976 Intersite Variability in the Mousterian of the Central Negev. In *Prehistory and Paleoenvironments in the Central Negev, Israel*, edited by A.E. Marks, pp. 113-140. SMU Press, Dallas.

National Museum of Kenya
1993 *Division of Archaeology Conditions of Affiliation.* Nairobi.

Nelson, Harry, and Robert Jurmain
1991 *Introduction to Physical Anthropology.* West Publishing Company, St. Paul, MN.

Newcomer, M.H
1975 Punch Technique and Upper Paleolithic Blades. In *Lithic Technology: Making and Using Stone Tools*, edited by E. Swanson, pp. 97-104. Mouton, Chicago.

Newcomer, M.H., R. Grace, and R. Unger-Hamilton
1986 Investigating Microwear Polishes with Blind Tests. *Journal of Archaeological Science* 13:203-207.

1988 Microwear Methodology: A Reply to Moss, Hurcombe, and Bamforth. *Journal of Archaeological Science* 15:25-33.

Odell, G. and F. Odell-Vereecken
1980 Verifying the Reliability of Lithic Use-Wear Assessments by "Blind Tests": The Low Power Approach. *Journal of Field Archaeology* 7:87-120.

Odell, George H.
1981 The Mechanics of Use-Wear Breakage of Stone Tools: Some Testable Hypotheses. *Journal of Field Archaeology* 8:197-209.

Odingo, Richard S.
1971a *The Kenya Highlands: Land Use and Agricultural Development.* East African Publishing House, Nairobi.

1971b Settlement and Rural Development in Kenya. In *Studies in East African Geography and Development*, edited by S. Ominde, pp. 162-176. University of California Press, Berkeley.

Ojany, Francis F.
1971 Drainage Evolution in Kenya. In *Studies in East African Geography and Development*, edited by S. Ominde, pp. 137-145. University of California Press, Berkeley.

Ominde, Simeon H.
1971 The Semi-Arid and Arid Lands of Kenya. In *Studies in East African Geography and Development*, edited by S. Ominde, pp. 146-162. University of California Press, Berkeley.

Owako, Frederick N.
1971 Machakos Land and Population Problems. In *Studies in East African Geography and Development*, edited by S. Ominde, pp. 177-193. University of California Press, Berkeley.

Patterson, L.W.
1979 Quantitative Characteristics of Debitage from Heat Treated Chert. *Plains Anthropologist* 24(85):255-260.

Phillipson, D.
1977 *The Later Prehistory of Eastern and Southern Africa.* Heinemann, London.

1993 *African Archaeology.* Cambridge University Press, Cambridge.

Purdy, Barbara A., and H.K. Brooks
1971   Thermal Alteration of Silica Minerals: An Archaeological Approach. *Science* 173:322-325.

Renne, P. R., W. D. Sharp, A.L. Deino, G. Orsi, and L. Civetta
1997 ^{40} Ar/^{39} Ar Dating into the Historical Realm: Calibration against Pliny the Younger. *Science* 277:1279-1280.

Rick, John W., and Sylvia Chappell
1983   Thermal Alteration of Silica Minerals in Technological and Functional Perspective. *Lithic Technology* 12:69-80.

Rightmire, G. Philip
1978   Human Skeletal Remains from the Southern Cape Province and Their Bearing on the Stone Age Prehistory of South Africa. *Quaternary Research* 9:219-230.

1979   Implications of Border Cave Skeletal Remains for Later Pleistocene Human Evolution. *Current Anthropology* 20(1):23-35.

1989   Middle Stone Age Humans from Eastern and Southern Africa. In *The Human Revolution: Behavioral and Biological Perspectives on the Origins of Modern Humans*, edited by P. Mellars and C. Stringer, pp.109-122. Princeton University Press, Princeton, NJ.

Robertshaw, Peter
1995   The Last 200,000 years (or Thereabouts) in Eastern Africa: Recent Archaeological Research. *Journal of Archaeological Research* 3(1):55-86.

Rolland, Nicolas, and Harold L. Dibble
1990   A New Synthesis of Middle Paleolithic Variability. *American Antiquity* 55(3):480-489.

Sackett, James R.
1973   Style, Function and Artifact Variability in Paleolithic Assemblages. In *The Explanation of Culture Change*, edited by Colin Renfrew, pp.317-25. University of Pittsburgh Press, Pittsburgh, PA.

1977   The Meaning of Style in Archaeology: A General Model. *American Antiquity* 42 (3):369-380.

1982   Approaches to Style in Lithic Archaeology. *Journal of Anthropological Archaeology* 1:59-112.

1986   Style, Function, and Assemblage Variability: A Reply to Binford. *American Antiquity* 51(3):628-634.

Sampson, C. Garth
1968   *The Middle Stone Age Industries of the Orange River Scheme Area*. National Museum, Bloemfontein Memoir No. 4. South Africa.

1974   *The Stone Age Archaeology of Southern Africa*. Academic Press, New York.

Schick, Kathy D., and Nicholas Toth
1993   *Making Silent Stones Speak*. Simon and Schuster, New York.

Schmitt, K.
1992   Anthropo-Zoogenic Impact on the Structure and Regeneration of a Submontane Forest in Kenya. In *Tropical Forests in Transition: Ecology of Natural and Anthropogenic Disturbance Processes*, edited by J.G. Goldammer, pp.105-126. Birkhäuser, Berlin.

Schweitzer, F.R.
1970   A Preliminary Report of Excavations of a New Cave at Die Kelders. *South African Archaeological Bulletin* 25:136-138.

Semaw, S., P. Renne, J.W.K. Harris, C.S. Felbel, R.L. Bernor, N. Fesseha, and K. Mowbray
1997   2.5 Million Year Old Stone Tools from Gona, Ethiopia. *Nature* 385:333-336.

Semenov, S.A.
1964   *Prehistoric Technology*. Barnes and Noble, London.

Shackleton, Nicholas J.
1982   Stratigraphy and Chronology of the KRM Deposits: Oxygen Isotope Evidence. In *The Middle Stone Age at Klasies River Mouth in South Africa*, edited by R. Singer and J. Wymer, pp.194-199. The University of Chicago Press, Chicago.

Shackleton, R. M.
1946   *The Geology of Nanyuki and Maralal Area*. Report No. 11. Government Printer, Nairobi.

Shafer, Harry J.
1978   *Lithic Technology at the George C. Davis Site, Cherokee County, Texas*. Ph.D. dissertation. The University of Texas, Austin.

Simmons, Alan H.
1982   Chipped Stone Glossary. In *Cholla Project Archaeology, vol. 1: Introduction and Special Studies*, edited by J. Jefferson Reid, pp. 35-45.

Arizona State Museum Archaeological Series No. 161. Tucson.

1983   Excavation at Prehistoric Sites on the Navajo Indian Irrigation Project, Northeastern New Mexico. *Journal of Field Archaeology* 10:155-176.

Singer, R., and J. Wymer
1982   *The Middle Stone Age at Klasies River Mouth in South Africa.* The University of Chicago Press, Chicago.

Soja, Edward W.
1968   *The Geography of Modernization in Kenya: A Spatial Analysis of Social, Economic, and Political Change.* Syracuse Geographical Series No. 2. Syracuse University Press, Syracuse, NY.

Sollberger, J.B.
1986   Lithic Fracture Analysis: A Better Way. *Lithic Technology* 15:101-105.

1994   Hinge Fracture Mechanics. *Lithic Technology* 19:17-20.

Speth, John
1972   Mechanical Basis of Percussion Flaking. *American Antiquity* 37:34-60.

1974   Experimental Investigations of Hard-hammer Percussion Flaking. *Tebiwa* 17:7-36.

Stafford, Barbara D.
1980   Prehistoric Manufacture and Utilization of Lithics from Corduroy Creek. In *Studies in the Prehistory of the Forestdale Region, Arizona,* edited by C. Russell Stafford and Glen E. Rice, pp. 251-297. Anthropological Field Studies No. 1. Arizona State University, Tempe.

Stoneking, Mark, and Rebecca L. Cann
1989   African Origin of Human Mitochondrial DNA. In *The Human Revolution: Behavioral and Biological Perspectives on the Origins of Modern Humans,* edited by P. Mellars and C. Stringer, pp.17-30. Princeton University Press, Princeton, NJ.

Stringer, C.B., and P. Andrew
1988   Genetic and Fossil Evidence for the Origin of Modern Humans. *Science* 239:1261-1268.

Stringer, C.B., and C. Gamble
1993   *In Search of the Neanderthals.* Thames and Hudson, London.

Sullivan, Alan P., III, and Kenneth C. Rozen
1985   Debitage Analysis and Archaeological Interpretation. *American Antiquity* 50:755-779.

Szabo, B.J., and K.W. Butzer
1979   Uranium-Series Dating of Lacustrine Limestones from Pan Deposits with Final Acheulian Assemblage at Rooidam, Kimberley District, South Africa. *Quaternary Research* 11:257-160.

Tarbuck, Edward J., and Frederick K. Lutgens
1993   *The Earth: An Introduction to Physical Geology.* Macmillan Publishing Company, New York.

Tringham, Ruth, Glenn Cooper, George Odell, Barbara Voytek, and Anne Whitman
1974   Experimentation in the Formation of Edge Damage: A New Approach to Lithic Analysis. *Journal of Field Archaeology* 1:171-196.

Tsirk, Are
1979   Regarding Fracture Initiations. In *Lithic Use-Wear Analysis,* edited by B.Hayden, pp. 83-96. Academic Press, New York.

Van Noten, Francis
1977   Excavations at Matupi Cave. *Antiquity* 51:35-40.

1982   *The Archaeology of Central Africa.* Akademische Druck-u. Verlagsanstalt, Graz, Germany.

Van Peer, Philip
1992   *The Levallois Reduction Strategy.* Prehistory Press, Madison, WI.

Vincent, C.E.,T.D. Davies, P. Brimblecombe, and A.K.D. Beresford
1989   Lake Levels and Glaciers: Indicators of the Changing Rainfall in the Mountains of East Africa. In *Quaternary and Environmental Research on East African Mountains,* edited by W.C. Mahaney, pp.199-216. York University, Toronto.

Vogel, J.C.
1970   Groningen Radiocarbon Dates IX. *Radiocarbon* 12(2):444-471.

Vogel, J.C., and P.B. Beaumont
1972   Revised Radiocarbon Chronology for the Stone Age in South Africa. *Nature* 237:50-51.

Volman, Thomas P.
1984   Early Prehistory of Southern Africa. In *Southern African Prehistory And*

Paleoenvironments, edited by R.G. Klein, pp.169-220. A.A. Balkema, Rotterdam.

Wayland, E.J.
1924   The Stone Age in Uganda. *Man* 24.

Wendorf, Fred, R.L. Laury, Romauld Schild, C. Vance Haynes, and Paul E. Damon
1975   Dates for the Middle Stone Age of East Africa. *Science* 187:740-742.

White, Roberts S., and Dan P. McKenzie
1989   Volcanism at Rifts. *Scientific American* 261 (1):62-71.

Whittaker, John C.
1994   *Flintknapping: Making and Understanding Stone Tools.* The University of Texas Press, Austin.

Wilmsen, Edwin N.
1967   *Lithic Analysis and Cultural Inference: A Paleo-Indian Case.* Ph.D. dissertation, University of Arizona, Tucson.

1968   Functional Analysis of Flaked Stone Artifacts. *American Antiquity* 33(2):156-161.

Winner, Langdon
1977   *Autonomous Technology: Technics-Out-of-Control as a Theme in Political Thought.* The MIT Press, Cambridge, MA.

Wolpoff, Milford H.
1989   Multiregional Evolution: The Fossil Alternative to Eden. In *The Human Revolution: Behavioral and Biological Perspectives on the Origins of Modern Humans*, edited by P. Mellars and C. Stringer, pp.62-108.   Princeton University Press, Princeton, NJ.

Wolpoff, Milford H., and Alan Thorne
1991   The Case against Eve. *New Scientist* 1774:37-41.

Yellen, J., A. Brooks, E. Cornelissen, M. Mehlman, and K. Stewart
1995   A Middle Stone Age Worked Bone Industry from Katanda, Upper Semliki Valley, Zaire. *Science* 268:553-556.

Yerkes, Richard W, and P. Nick Kardulias
1993   Recent Developments in the Analysis of Lithic Artifacts. *Journal of Archaeological Research* 1(2):89-119.

Appendix A
Paleolithic in East Africa

In general, the African Stone Age can be divided into three parts: Earlier, Middle, and Later Stone Ages. This scheme was first suggested by Goodwin and van Riet Lowe (1929) on the basis of archaeological material found from South Africa. At first these three ages were separated from each other by two Intermediate periods. The Middle Stone Age was separated from the Earlier Stone Age by the First Intermediate Period and from the Later Stone Age by the Second Intermediate Period which is characterized by the coexistence of making the Levallois or disc core and the appearance of microliths. These two intermediate periods were used to cover the transitional periods between the end and beginning of the Age.

Although the framework of classification was largely accepted in most of sub-Saharan Africa, typology and nomenclature generated a debate. Some scholars such as Leakey (1931) preferred to use European terms, others (Goodwin 1926) suggested dropping European terminology in favor of local terminology. As a result, a Wenner-Gren symposium on African prehistory was held at Burg-Wartenstein in 1965. This symposium recommended abandonment of the age system and the intermediate periods in Africa because the terms have both cultural and chronological connotations and a hierarchical scheme was suggested to replace the age system (Bishop and Clark 1967:892-897). Despite the problem of terminology, the age system is still used by most scholars (Clark 1970) because it conveniently labels both age and order. However, the two intermediate periods as well as various terms of Middle Stone Age named after local industries, hamper the study of African prehistory. In order to avoid the confusion caused by so many different terms, the three age system suggested by Goodwin and van Riet Lowe is used in the present study.

Conventionally, the African Stone Age is regarded as synonymous with the Eurasian Lower, Middle, and Upper Paleolithic periods, although the former cannot be fully correlated with the latter. The Lower Paleolithic ends with the Acheulian, while the Earlier Stone Age survives in Africa considerably later than that. The Middle Paleolithic is closely associated with the Mousterian flake industries, whereas the early MSA culture is related to Mousteroid industries and some of the later MSA culture is similar to the Upper Paleolithic culture of Europe. In addition, part of the Later Stone Age in Africa overlaps the Mesolithic of Eurasia (Phillipson 1993:60, 61) (Table 29).

The Earlier Stone Age begins with first appearance of the stone artifacts and ends with the disappearance of large cutting tools such as handaxes and cleavers. The first stone artifacts were found at Kada Gona and the adjacent Gona River drainage in the Hadar region of Ethiopia (Semaw *et al.* 1997). The underlying volcanic tuff of the former site was dated by potassium-argon and fission-track techniques. This dating method suggested 2.8 " 0.2 million years for the tuff. As the 18 stone artifacts recovered from Kada Gona were believed to be slightly younger than the volcanic tuff, the age put on the artifacts was between 2.5 and 2.7 million years (Lewin 1981:807). In the latter case, radio-isotopic age control and a magnetic polarity stratigraphy suggest that the stone artifacts were 2.6-2.5 million years old. The stone artifact assemblages of the early Earlier Stone Age usually consist of pebbles, cobbles, slabs, blocks, intentionally produced flakes and angular fragments. These stone artifacts are characterized by informal trimming and emphasis on edge modification. This variety of primary forms began to be changes around 1.6 to 1.5 million years ago. From this period, the African stone artifact assemblage shows an increase in retouched stone artifact forms such as handaxes and cleavers.

The following Middle Stone Age is characterized by a variety of prepared cores and retouched flakes. Since the break between the Earlier and the Middle Stone Ages was not a sharp one, dating the end of the Earlier Stone Age and the beginning of the Middle Stone Age is problematic in African archaeology (Volman 1984:170, 172). This dating problem will be discussed in the following section.

The Later Stone Age is dominated by microlithic stone artifacts, which lack the characteristic Middle Stone Age retouched tools, flakes, and cores. The Later Stone Age is also characterized by other elements of material culture such as rock painting, formal burial, and so on. The items of material culture are made and used by modern African hunter-gatherers as well (Deacon 1984:221; Volman 1984:174).

Middle Stone Age in East Africa

The Middle Stone Age is separated from Earlier Stone Age by the First Intermediate Period and from the Later Stone Age by the Second Intermediate Period. However, the Intermediate Periods are not clearly distinguished from the Middle Stone Age and industries of these periods gradually merged into one another.

Dates of the Middle Stone Age

In studying late Pleistocene, one of the most problematic issues is defining an archaeological specimen that belongs to this period. In archaeology generally, the most reliable chronometric dating methods are obtained by radiocarbon and potassium-argon techniques. Unfortunately, these two methods are not very useful in dating the archaeological material of this period. Although potassium-argon dating can be theoretically used to date igneous rock ranging in time from the formation of the Earth to as recently as 10,000 years B.P., the half-life of potassium is so great that igneous rocks less than about 500,000 years of age can be dated only if they are especially rich in that element. Furthermore, potassium-rich rocks less than 100,000 years old are extremely hard to date. As a result, the potassium-argon dating technique is useful to date only material that is older than 500,000 years (but cf., Renne *et al.* 1997). In contrast, radiocarbon technique is used to date material which is 50,000 years or younger. Materials greater than

Table 29. Chronology of Paleolithic Period in Eurasia and Stone Age in Africa

Time (B.P)	Paleolithic Time	Stone Age	Time (B.P)
10,000	Neolithic Mesolithic	Neolithic	3,000 (?)
35,000	Upper Paleolithic	Later Stone Age	40,000
90,000	Middle Paleolithic	Middle Stone Age	200,000
2,900,000	Lower Paleolithic	Earlier Stone Age	,900,000

Modified from Dickson (1993b:Figure 7).

40,000 years in age must be artificially enriched in order to be dated. However, the use enrichment process can date only specimens which are younger than 75,000 years old (Bennett 1979). That is, dating of the Middle Stone Age remains problematic since it falls between the ranges of radiocarbon and potassium-argon techniques.

In addition, many MSA sites that have yielded human remains have produced an admixture of older and younger materials due to stratigraphic or disturbance problems (Brooks et al 1990:61). As a result, there is little consensus as to the nature of the period of the Middle Stone Age (Bordes 1968; Clark 1970; Phillipson 1993; Robertshaw 1995).

Dating the Middle Stone Age in East Africa. In order to define the age of MSA a variety of dating methods have been used in the archaeological sites in sub-Saharan Africa. At Porc Epic Cave in east central Ethiopia obsidian hydration dating of the MSA component wasattempted. The minimum age estimates of the site indicate that Proc Epic Cave was occupied between around 61,000 B.P. and 77,500 B.P. (Michels and Marean 1984:68).

Michels, Tsong, and Nelson (1983:363-364) applied obsidian hydration dating at the Prospect Farm MSA site in the Naivasha-Nakuru basin region of Kenya. The site is composed of four main episodes of occupation: Middle

Stone Age, Second Intermediate Period or early Later Stone Age, the Eburran Industry, and the Pastoral Neolithic. The earliest episode of occupation is dated to 119,646 ± 1668 yr B.P., while the youngest MSA 50,777 ± 3322 yr B.P. The transition between MSA and Second Intermediate technologies appeared between 46,700 and 53,600 years ago. Thus, Michels et al. suggest that the site was occupied around 120,000 years ago and that the beginning of the Second Intermediate in East Africa falls between 45,000 and 50,000 years ago. Hydration dates from the site also suggest that Second Intermediate or Early Later Stone Age appeared between 21,800 and 32,500 years ago.

Brooks et al. (1995) dated the MSA site of Katanda at the Upper Semliki Valley, Zaire using electron spin resonance, thermoluminescence, uranium series, and amino acid racemization. According to their results, dates of MSA site ranges from 89,000 to 160,000 B.P. These dates were consistent with a date obtained by faunal and stratigraphic data. Brooks et al.'s MSA dating overlaps the age obtained from the MSA site of Mumba in Tanzania. Based on uranium decay and amino acid racemization dates, Mumba site was occupied between 40,000 and 130,000 years ago (Mehlman 1987:133).

The oldest MSA date in East Africa comes from Gademotta in the Main Rift Valley of central Ethiopia. Potassium-argon dating technique applied to volcanic ashfall overlying MSA deposits yielded ages of 181,000 " 6,000 B.P. and 149,000 " 13,000. This date suggests that the MSA of East Africa began prior to 180,000 years ago (Wendorf et al. 1975).

Dating the Middle Stone Age in South Africa. Many analyses of dating MSA come from South Africa. Brooks et al. (1990) used protein diagenesis of ostrich eggshell to date an open-air pan-margin site, …Gi in the northwest Kalahari desert, on the Botswana-Namibia border. The site consists of three main cultural components: the MSA, with finely made bifacial and unifacial points and scrapers, an intermediate industry, and the LSA. The MSA component dates to between 65,000 and 85,000 years. The age obtained by protein diagenesis in ostrich eggshell almost overlaps the MSA date obtained from thermoluminescence technique. The latter method suggested 77,000 "11,000 B.P. for the MSA component of the site. A piece of eggshell produced from the base of unit 2C which belonged to Intermediate or the early post-MSA provided a radiocarbon age of 34,000 " 11,000 B.P.

Perhaps the most famous MSA site in Africa is Klasies River Mouth (KRM) caves and rockshelters on the Tzitzikama coast of Southern Cape Province, South Africa. These sites contain a long sequence of MSA horizons labeled 1, 1A, 1B, 1C, and 2. While radiocarbon dating of the cave I was more than 30,000 yr B.P., the more questionable technique of the amino acid racemization reaction study of bone samples from the cave I suggested ages of 65,000 to 90,000 years for the Sands-Ash-Shell (SAS) member. The SAS member is composed of finer sands containing shell middens and other remains of human occupation. The SAS is overlaid by the Light-Brown Sands (LBS) member, which produced bones and was base of the MSA deposits. This LBS member was dated 110,000 years (Bada and Deems 1975). Electron-Spin-Resonance (ESR) analysis of tooth enamel suggested age of 93,500 ± 10,400 and 88,300 ± 7,800 years (Grün, Shackleton, and Deacon 1990:429) which was consistent with the age obtained by the amino-acid racemization dating.

Butzer (1982:41-42) correlated KRM Cave I deposit with the sea level stratigraphy, especially with that of the last interglacial period. The radiometric dating of fossil corals has suggested that the last interglacial was 120,000 yr B.P. (Ku *et al.* 1974:959-962). Shackleton compared and correlated the fragmentary oxygen isotope record contained in a shell-midden sequence with the continuous record obtained from deep-sea cores. According to him (Shackleton 1982:196-199), the MSA I middens of KRM can only have formed during isotope substage 5e and MSA II probably either substage 5c or 5a. Isotope substage 5e is the last time when the sea was isotopically as light as it is today. During substage 5e of interglaciation the sea was about 6-8 m above its present level. It has been widely accepted that this high sea level event occurred about 120,000 yr B.P. Shackleton suggested that if substage 5e consisted of two substages, the KRM MSA I could be dated between 130,000 and 120,000 years B.P.

However, there is disagreement about the timing of the high sea level. The KRM caves reveal that they were cut by a sea level between four and 20 meters; KRM Cave I by six to eight meters. Based on the study of the late Cenozoic high sea levels, Hendey and Volman (1986) suggest that the 6-8 m shoreline in South Africa occurred in the early Pleistocene, not the later Pleistocene. According to them, the four-meter shoreline appeared in the late Pleistocene correlated with isotope substage 5e. Thus, Hendey and Volman extend the beginning of the MSA at KRM further back in time. The final KRM MSA is recorded in layer 13, which is correlated with isotope stage four (ca. 70,000 B.P.). After this final MSA occurrence, the site was apparently remained unoccupied between 65,000 and 5,000 B.P. (Butzer 1982:35-41) (Table 30).

The Rooidam Site near Kimberley, South Africa produced late Acheulian artifacts. Lacustrine limestone samples from sedimentary pan deposits that overlaid the archaeological artifacts were dated using the uranium series dating technique. The oldest date obtained from the samples was 174,000 " 20,000 years old.

As prior to this phase, Acheulian settlement focused on a shore zone of shallow lakes, a minimum age of about 200,000 yr B.P. was assigned to the terminal Acheulian (Szabo and Butzer 1979:257, 260). This terminal Acheulian date is correlated with date of Border Cave, KwaZulu, South Africa. Border Cave is famous not only for MSA sequence and associated hominid, but also for the earliest demonstrable LSA strata in southern Africa. Lithostratigraphic and sedimentological study suggest that the cave contains eight Pleistocene sedimentary cycles, including six major cold phases and w intervening weathering horizons. The six cold phases were correlated with isotope stage 4, 5, and 6. This correlation places the MSA deposit at ca.195,000 yr B.P. The youngest MSA deposit dates range from 45,400 to greater than 49,100 yr B.P., while the earliest datable LSA age falls between 35,700 to 37,500 yr B.P. (Butzer, Beaumont, and Vogel 1978). In addition to the foregoing sites described above, there are a number of sites that apparently contain final MSA components. One of these sites is Die Kelders Cave I site on the Walker Bay coast in South Africa. The MSA sequence at Die Kelders is correlated with all or parts of isotope stages 4 and 3 which range between 30,000 and 50,000 years ago (Klein 1975; Hendey and Volman 1986:196).

Bushman Rockshelter and Heuningneskrans Cave are located in the eastern Transvaal in South Africa. The mixed MSA and LSA deposit of the Bushman Rockshelter was radiocarbon dated. The oldest date returned was 31,900 yr B.P. (Butzer 1984:59-61). The Heuningneskrans Cave was radiocarbon dated to 50,000 B.P. for the early LSA deposit (Vogel and Beaumont 1972).

Boomplaas Cave above the Congo Valley in South Africa contains a sequence from the Upper Pleistocene to the Holocene. Radiocarbon dates suggest that the oldest deposit at the site is 80,000 B.P. with the most intensively focused period ranging in age from 60,000 to 18,000 B.P. The youngest MSA occurrence was between 40,000 and 30,000 B.P. (Deacon 1979:246-254).

In terms of the cultural sequence, the two intermediate periods in sub-Saharan African archaeology are rarely used today. These two periods are usually included in either MSA or LSA. In the present study, the Second Intermediate is included in the final MSA. Based upon dates of MSA obtained from various dating methods described above, it is reasonable to suggest that the MSA period dates from ca. 200,000 yr B.P. That is, the MSA begins with the appearance of the prepared cores and flake tools rather than core tools and lasts until around 40,000 B.P. when the microlithic technology appears (Robertshaw 1995:60; Volman 1984:184). In fact, many final MSA/LSA sites in South Africa reveal that early LSA intruded into South Africa shortly before 50,000 B.P. The end date for the MSA has been obtained by radiocarbon dating method. As the effective range of the Radiocarbon method is 40,000 B.P., the suggested final MSA is minimum dates of the final MSA and it can be extended further back (Vogel and Beaumont 1972:51). Table 31 summarizes the various MSA dates and sites in sub-Saharan Africa. Table 32 shows Plio-Pleistocene and Paleolithic time scale in Africa.

Table 30. Oxygen Isotope Stage

Approximate years B.P.	Oxygen isotope stage	Inferred worldwide climate		Geologic time
0				
	1	Very warm		Holocene
12,000				
	2	Very Cold	Last Glacial	Upper Pleistocene
	3	Cold with warm oscillations		
64,000				
	4	Very cold		
75,000				
C. 82,000	5a	Warm	Last Interglacial	
	5b	Cold		
C. 105,000	5c	Warm		
	5d	Cold		
	5e	Very warm		
128,000				
	6	Cold with warm oscillations		Middle Pleistocene
195,000				
	7	Warm with cold oscillations		
251,000				
730,000	19			
1,700,000				Lower Pleistocene
				Pliocene

Adapted from Volman (1984:Table 1).

Hominids of the Middle Stone Age

It is important to know the nature of the hominids that were responsible for producing the archaeological remains of the Middle Stone Age. This is because the evolutionary change, morphology, intelligence, and hominid behavior are no doubts related to the changes in the MSA technology/industry through time. Furthermore, some of the MSA archaeological record is believed to be the expression of the behavior of the earliest anatomically modern humans (*Homo sapiens sapiens*). Morphologically, anatomically modern humans have shorter jaws than archaic *Homo sapiens* and the former exhibit more developed chins and smaller teeth than the latter. *Homo sapiens sapiens* has skull forms that are characterized by small or no brow ridge and shorter and high skull. Their brain size ranges from 1,200 to 1,700 ml and they are, in general, taller (Jones, Martin, and Pilbeam

1992:251). Although there are numerous ways of defining modern humans, they are beyond the scope of the present study.

The hominids of the Middle and early Upper Pleistocene are poorly known. As a result, the issue question of the origins of anatomically modern humans attracts many researchers to the African MSA. Until recent, it has not been certain what hominids were responsible for the early post-Acheulian industries.

Africa's Eve and the Rapid Displacement Model. The origins of anatomically modern human became an issue of hot debate following some provocative studies of mitochondrial DNA (mtDNA). According to Cann *et al.* (1987), mtDNA is transmitted strictly through the maternal line. If any variant appears in a group of lineages, it is interpreted to mean that a mutation occurred in the ancestral lineage. In addition, it is assumed that each individual is homogeneous for its multiple mtDNA genomes. Therefore, a tree diagram related to types of human mtDNA can also be considered a genealogy linking maternal lineages in modern human population to a common ancestral female. This mtDNA study suggested two important conclusions: first, that Africa is the source of the modern human mitochondrial gene pool and second, that the common ancestor of all mtDNA types existed 140,000-290,000 years ago. This time scale is obtained from mtDNA sequence divergence accumulated at a constant rate in human based on the time required for the peopling on New Guinea, Australia, and the New World. Using these peopling rates the researchers were able to calculate the mean rate of mtDNA divergence within humans, which occurs two and four percent per million years. The analysis of mtDNA shows that the common ancestral mtDNA links mtDNA types that have diverged by an average of nearly 0.57 percent. As a result, the calculated time is 140,000-290,000 years ago. Recent mtDNA study by Stoneking and Cann (1989:17-28) extends the existence of the common ancestor of all modern humans to between 50,000 and 500,000 years ago. Within 100,000 years, so the argument goes, anatomically modern humans migrated out of Africa and eventually replaced the indigenous human populations throughout the world. This replacement was achieved without population admixtures between the indigenous and new populations (Stringer and Andrew 1988).

Multiregional Model. Alternative to the African Eve model, the multiregional model is based on the human fossil records and suggests that anatomically modern human evolved from the earlier archaic *Homo sapiens* populations in a number of different regions of the world. This model proposes that the common ancestor of all modern human population was Homo *erectus* who first migrated out of Africa around 850,000 years go. As anatomically modern humans evolved from *Homo erectus* and did not suddenly appear, there is no need to seek a single speciation event. Instead, this model postulates that

morphological differentiation occurred as local evolution at many times and places in the Old World (Wolpoff 1989; Wolpoff and Thorne 1991).

Archaeological Evidence. The MSA is important in this question because it overlaps in the key time period when modern humans appeared. Unfortunately, only a few MSA archaeological sites in East Africa have produced human remains and most of these have yielded only incomplete individuals or fragments. The oldest archaeological evidence of anatomically modern human comes from the southern part of Africa including the MSA sites of Klasies River Mouth Cave, Border Cave, and Florisbad Site.

Bits of cranium, mandible, isolated teeth, and a few broken postcranial items were recovered from Klasies River Mouth (KRM), South Africa. Except for the layers associated with the Howieson's Poort Industry, the whole sequence from MSA I to MSA III at the site produced human remains. Many of the skeletal remains appear to be anatomically modern. Most notably, a frontal bone fragment with part of supraorbital ridge and nasal bone (specimen number 16425: KRM 1-16 from MSA II) shows slight supraorbital development which is characteristic of the crania of many modern Africans such as the modern Bushman. Many other skeletal remains are recovered from MSA layers. These also have a modern morphological appearance. The most complete mandible found at KRM comes from MSA I deposit. This mandible consists of right premolar and first molar and left first molar and second molar (specimen number 41815: KRM 1B-10). This left mandible bone does not indicate any archaic morphology in the facial skeleton. As a result, the KRM human remains have been classed as anatomically modern humans (*Homo sapiens sapiens*) (Rightmire 1978:219, 221; Singer and Wymer 1982). As mentioned above, the earliest KRM MSA I is dated 120,000-130,000 B.P. and KRM MSA II 100,000-110,000 B.P (Bada and Deems 1975; Butzer 1982; Shackleton 1982). These dates make the KRM human remains the earliest known examples of modern human yet recovered.

Border Cave, situated on the boundary between Swaziland and South Africa yielded human remains including an incomplete mandible, partial adult cranium, and an infant skeleton. Of these, the most interesting is the partial cranium. This cranium includes most of the frontal bones, parts of both parietals and temporals, an occipital fragment, and a right zygomatic bone. Recent statistical analysis comparing frontal form, supraorbital development of this cranium with modern Bushmen, Zulu, and Sotho crania suggests the Border Cave hominid is directly linked to modern African populations (Rightmire 1979). Based on lithostratigraphic and sedimentological correlations with the deep-sea isotope stages, the mandible and cranium have been dated to 115,000 B.P., the infant skeleton to 90-95,000 B.P. (Butzer 1979:28). However, the ages of hominid fossils remain controversial in part because of the poor controls exercised in excavating them. To settle this debate, it will necessarily to date the fossils themselves.

Table 31. Dates of Middle Stone Age in Sub-Saharan Africa*

East Africa	Site	Country	Date (B.P.)	Method
	Porc Epic Cave	Ethiopia	61,000 ~ 77,500	Obsidian hydration
	Prospect Farm	Kenya	45,000 ~ 120,000	Obsidian hydration
	Katanda	Zaire	89,000 ~ 160,000	Electron-Spin-Resonance Thermoluminescence Uranium series decay Amino acid racemization
	Mumba	Tanzania	40,000 ~ 130,000	Amino acid racemization Uranium series decay
	Gademotta	Ethiopia	180,000	Potassium-argon
South Africa	▢ Gi	Botswana	65,000 ~85,000	Protein diagenesis in ostrich eggshell
	Heuningenes-krans	South Africa	>50,000	Radiocarbon
	Bushman Rockshelter	South Africa	30,000 ~ 50,000	Radiocarbon
	Boomplaas Cave	South Africa	30,000 ~ 80,000	Radiocarbon
	Die Kelders Cave	South Africa	30,000/50,000 ~ 120,000	Oxygen isotope record
	Klasies River Mouth Cave	South Africa	65,000 ~ 120,000/130,000	Amino acid racemization Electron-Spin-Resonance Oxygen isotope record Stratigraphic analysis
	Rooidam	South Africa	190,000	Uranium series decay
	Border Cave	South Africa	45,400/49,100 ~ 195,000	Lithostratigraphic and sedimentologic study

* More information about MSA sites in South Africa is provided by Volman (1984), Butzer (1984), and Singer and Wymer (1982). This table summarizes oldest known MSA sites with final MSA / early LSA and their dates in both East and South Africa.

Table 32. Geological and Archaeological Time Scale

Time (B.P.)	Geological Time	Archaeological Time	Time (B.P.)
10,000	Holocene	Neolithic	3,000 (?)
		(Upper Paleolithic)	(35,000)
	Upper Pleistocene	Later Stone Age	40,000
128,000		(Middle Paleolithic)	(90,000)
	Middle Pleistocene	Middle Stone Age	
			200,000
730,000		Earlier Stone Age (Lower Paleolithic)	
	Lower Pleistocene		
1,720,000			
	Late Pliocene		
			2,900,000 (2,900,000)
3,400,000			

Modified from Dickson (1993b:Figure 7).

The Florisbad Site in the Orange Free State, South Africa is one of the oldest known MSA sites on the continent. The site, first excavated in 1932, produced a human cranium and became the type assemblage for MSA fauna in southern Africa. In the 1980s numerous unanswered questions about the site lead to its re-excavation. A major question concerned the affinities and age of the human remains. Before this re-investigation, the human cranium was attributed to anatomically modern human and assigned to MSA. The new investigation revealed that the human remains discovered at the site belongs to an archaic *Homo sapiens* although the cranium looks more modern than those human fossils produced by Broken Hill in Zambia and Saldanha in South Africa. Based on anatomical comparison and the supposed age of the Florisian fauna, the estimated age of the Florisbad cranium is 100,000-200,000 years (Kuman and Clarke 1986).

In Eastern Africa, the most important MSA sites to have produced hominid remains are Omo in Ethiopia and Laetoli in Tanzania. The Omo site produced the remains of three adult individuals, designated Omo I, Omo II, and Omo III all of whom were recovered from the Kibish Formation. Based on the shells in the deposit, the suggested date of Omo Site is 130,000 yr B.P. (Butzer 1987; Leakey, Butzer, and Day 1969).

The Ngaloba beds at Laetoli have also yielded a hominid fossil cranium (LH 18). The LH 18 cranium is composed of an almost complete and intact cranial vault, much of the base, both temporal bones, part of the sphenoid bone, and fragment maxillo-facial bone. Based on the tuff in the lower unit of the Ndutu beds at Olduvai Gorge which was potassium-argon dated, the estimated age of the Ngaloba beds is 120,000 ± 30,000 yr B.P. (Day, Leakey, and Magori 1980).

The taxonomic status of the human remains discovered from these two sites still need to be confirmed (Nelson and Jurmain 1991:534). The fossil remains from these two sites have variously been classed as transitional forms of *Homo erectus/Homo sapiens*, archaic *Homo sapiens*, Homo *sapiens,* or *Homo sapiens sapiens* (Day, Leakey, and Magori 1980; Leakey, Butzer, and Day 1969; Rightmire 1989). Thus, at the present time, the only remains that have been confirmed as anatomically modern humans are those associated with MSA at the Klasies River Mouth Site.

Apart from the fossil evidence, the timing of these origins of the anatomically modern human is also an important and controversial research question in contemporary paleoanthropology (Hayden 1993). In general, two classes of archaeological evidence have been used as indicators of the appearance of modern behavior: changes in human anatomy reflected in the fossil record and changes in the nature of human artifacts (Klein 1989; Mellars 1989; Stringer and Gamble 1993). Using the latter category of evidence, two alternative hypotheses about the dates for the development of modern behavior have recently been suggested. Recent excavations at three MSA archaeological sites in the Semliki Valley in Zaire have challenged our conception of the MSA. The lithic industry from Katanda 2 is characterized by numerous discoidal cores, few formal tools, and a single large fusiform bone point. The stone artifact assemblage from Katanda 9 includes retouched pieces, numerous discoidal cores with single and multiple platforms, and rare formal tools. The points or daggers and blades are rare. However, this site also contains seven well-made uniserial barbed bone points, two unbarbed points, and a large dagger-shaped object of unknown function. Katanda 16 also produces very similar artifacts to those from site 9 (Yellen et al. 1995:553-555). On the basis of their work in Zaire, Brooks *et al.* (1995) and Yellen *et al.* (1995) assert that sophisticated tools (and by implication, modern behavior) appeared in African prehistory during the MSA at a date much earlier than elsewhere in the world. Based on the thermoluminescence, which provides an age for quartz grains in the soil, and electron spin resonance methods, they argue that the appearance of sophisticated modern behavior can be back to at least 90,000 years ago. Arguing from both the hominid fossil record and stone artifact assemblages, Klein (1995) and many others propose that the development of modern behavior appeared in Africa no earlier than the Later Stone Age, about 50,000 years ago. As a result of the excavation of these sophisticated bone points, dates for the origins of modern human behavior have been questioned.

Defining Middle Stone Age in Africa

As noted, the three stone age systems of the African Paleolithic are not clear-cut divisions. Rather they slightly overlapped each other and exhibit a strong continuity between one another. Thus, it is necessary to discuss the post Acheulian times in order to define the MSA period. Like human fossil evidence, most information about MSA comes from South Africa and the MSA period in South Africa has been well established. It is appropriate, therefore, to discuss the MSA of South Africa first.

The Later or Terminal Earlier Stone Age in Southern Africa. In post-Acheulian times, Africa falls into two distinct zones: one north and one south of the Sahara. The former zone is linked with Europe and the Middle East and contains a Mousterian occupation in the strict sense. The latter southern zone exhibits lithic industries which are "Mousteroid" in character but chronologically roughly contemporary with the European Upper Paleolithic. One of the distinct characteristics of African post-Acheulian lithic industries is their great diversification. Numerous local facies emerge and spread over vast areas as humans began to occupy regions formerly uninhabited (Bordes 1970:121).

In post-Acheulian times, stone tools show the beginnings of (1) regional specialization, (2) the proliferation in the number of variable toolkits within the regions, and (3) a remarkable increase in the number of standardized forms of tools. This post-Acheulian diversification has been called the First Intermediate period (Clark 1970:108). Industries of this period are called various terms, such as Fauresmith, Sangoan, and Charaman; the terminology itself hinders the study of this period (Sampson 1974:103, 108). The term "First Intermediate" has been abandoned and most of the industries belong to this period have been lumped into the Earlier Stone Age, particularly the later Earlier Stone Age. Although the present study does not use these various local terms and eliminates these terms, it is necessary to understand them merge into one another. Therefore, in the following section, I use original terms to describe the different local industries in the final Earlier Stone Age.

Fauresmith Industry. Until recently, it was suggested that the Fauresmith Industry evolved from the local Acheulian tradition in the southern part of Limpopo because this type of material had not been discovered in the northern part of Limpopo. This tradition was assumed to have been used by peoples exploiting the montane grasslands, acacia grasslands, and dry steppe of the region. Typical Fauresmith tools include small hand axes, cleavers, and many tools made on flakes such as scrapers and chisels (Clark 1970:110). However, it has since been determined that the Fauresmith industry dates to the Acheulian times rather than the post-Acheulian times. Now the Fauresmith is seen as equivalent to the final Acheulian south of the Limpopo (Sampson 1974:108, 121, 187). The Rooidam Site (Fock 1968; Szabo and Butzer 1989) and Wonderwreck Cave (Malan and Cook 1941) are two well known Fauresmith sites or the final Acheulian Period in South Africa.

Sangoan and Charaman Industries. The study of the post-Acheulian archaeology of the region north of the Limpopo has focused on Zimbabwe and southwestern Zambia. The archaeological stone artifacts discovered from this northern part are found immediately above the final local Acheulian at certain key sites (Phillipson 1993:68). The stone-tool assemblages of these sites are formally differentiated from those of the final Acheulian by the presence of crude triangular-sectioned picks, thick core-axes and a variety of small flakes, especially small scrapers. These assemblages were attributed to the Sangoan, an industry typified by assemblages with a few crude-hand-axes and cleavers but abundant choppers, picks, and core axes (Sampson 1974:137, 138). Archaeological samples which were first designated as Sangoan industry were discovered by Wayland in 1920 based on his work at the type-site at Sango Bay in Uganda on the western shore of Lake Victoria (Wayland 1924). At first, prehistorians interpreted the tools in the Sangoan industry as representing a degeneration in ability and, by implication in the intelligence, of the makers. But Clark (1970:112-113) suggests that this change in the implements reflects the emergence of a forest adaptation and the corresponding need for some heavy-duty artifacts for intensive work in wood. Nearly all the large bifacial tools in the Sangoan industry have almost 100 percent of their perimeters retouched (McBrearty 1988:413). Clearly these tools are far from crude or unrefined. However, according to McBrearty (1990), plant macrofossils and faunal remains came from the Sangoan at the Simbi site in western Kenya indicate that the environment there was semiarid at the time of Sangoan occupation. Based on the Sangoan artifacts from the Muguruk site in western Kenya, McBrearty (1988:412-413) suggests that Sangoan artifacts might have been used for digging, or for some other task requiring a pointed object, rather than for wood working.

Sangoan materials discovered from more open plateau areas in southern Rhodesia and Zambia exhibit assemblages dominated by the light-duty flake tools such as points and scrapers rather than the heavy core-tools of the more northern region. These assemblages were first designated as the Charaman industry. The Charaman industry is now believed to be a later phase of Sangoan industry (Sampson 1974: 139-140). Many obstacles to the detailed study of Sangoan and Charaman industries remain. First, the unmixed unsorted assemblages from stratified datable contexts are too small to establish the accurate typological differences, second, associated faunal material is rare, and finally, there is an absence of unselected lithic samples (Phillipson 1993:68). The well-known sites of this later phase include Bambata and Pomongwe caves (Cooke 1963, 1969).

Transition from the Earlier Stone Age to the Middle Stone Age. The Earlier Stone Age was followed by complex of variant cultures that collectively belong to the Middle Stone Age. This transition from the ESA to the MSA has been framed in various ways. Some view this transition as a result of external cultural influences such as migration of new populations and the diffusion of new ideas into Africa (Goodwin and van Riet Lowe 1929:98), some emphasize the interplay between environmental and demographic and cultural factors as causal (Clark 1970:108), others credit unbroken evolutionary development within the of local industry (Mason 1962:156). Presently none of these explanations of the cause of the transition from the Earlier to Middle Stone Age has been universally accepted. Perhaps this is due to the lack of empirical data such as faunal and floral information and to the small artifact samples recovered from the period (Clark 1970:108-117; Deacon 1975:565; Phillipson 1993:60).

Middle and Later Stone Age in Southern Africa. As mentioned above, Goodwin first divided the prehistoric period in South Africa into the Earlier Stone Age and the Later Stone Age respectively. Later he found that these terms did not adequately cover the prehistoric period so he divided it into three ages on the basis of morphological features of local materials. Goodwin (1929:97-101, 143) defines that the Middle Stone Age as characterized by the preparation of cores, faceted striking platform of flakes, a variety of worked points, the dominance of flake tools, and the absence of true core tools such as hand-axe and cleavers that characterize the Earlier Stone Age. Although these artifact classes generally occur in most MSA assemblages through time, MSA assemblages came to be characterized by considerable local variability. With the improvement of technology and preference of raw material a particular stage of MSA yields different artifact morphology, core types, and raw materials. These differences between assemblages have lead to different local terminology for local MSA variants such as Pietersburg, Bambata, Howieson's Poort, Lupemban, and so on.

Pietersburg Complex Industry. This industry appears only to the south of Limpopo. The Pietersburg industry is typified by large number of blades and flake-blades, which are used directly without retouch; the elaborate secondary retouch of flakes to form uniform tool shapes is almost absent. As a result, the only formal tool types reported include scrapers and burins. Discoid, Levallois, and blade cores are generally present but the dominant core types vary from site to site. For instance, the Cave of Hearts produces many discoid cores, while Klasies River Mouth caves produced chiefly prepared or Levallois cores. The Pietersburg complex rarely produces triangular or convergent flake forms. In its later manifestations several morphological changes appear: large blade production decreases, while convergent flake forms and broader and smaller blades increase. Overall artifacts size decreases through time (Mason 1962; Sampson 1968:11-13; Singer and Wymer 1982). MSA I-II of Klasies River Mouth caves (Singer and Wymer 1982), Beds 1-5 of Cave of Hearts, Makapansgat (Mason 1962), Oringia I (Sampson 1968), and Layer C of Mossel Bay (Keller 1969) produces stone assemblages belonging the Pietersburg complex industry.

North of the Limpopo, the situation is different from that south of it. The northern woodland-savannah zone of Rhodesia and Zambia continues to be dominated by Sangoan-Charaman tradition of the MSA. Flaking technology based on the Levallois, discoid, and blade core techniques is common there but the production of blades is rare (Phillipson 1993:68). The Sangoan-Charaman industry is followed by Lupemban industry, which occurs, in general, in the central and eastern parts of Africa.

Bambata Complex Industry. Although this complex industry contains all the tool types, cores, and very similar techniques found in the Pietersburg complex industry, it exhibits several important features that were never found in the latter. The margins of flakes and blades of Bambata Complex are convergently trimmed to produce a point. Margins of flakes, blades, and blade fragments are elaborately retouched to be used as scrapers. Compared with Pietersburg industry, side scrapers in the Bambata complex are more numerous and show a wider variation in shape and size. Tools like borers, *outlis écaillés*, and grindstones are also found but the proportion of these formal tools in the assemblage vary (Sampson 1974:189, 191). The Bambata industry was produced by Beds 6-8 of Cave of Hearts, Beds II-III of Mwulu's Cave (Mason 1962), Bed KRM MSA I-III (Singer and Wymer 1982), and Middle Cave Earth of Border Cave (Cooke et al. 1945:8). The Bambata complex industry is referred to by numerous other names including Upper Pietersburg (Mason 1962), Normal Pietersburg (Cooke, Malan, and Wells 1945), and Stillbay (Goodwin and van Riet Lowe 1929). However, the term "Bambata" industry has not been applied to the extreme southern and interior parts of South Africa. Finally, the chronological relation with the Pietersburg industry has still remained unsolved. However, some sites yield that Bambata industry overlies the Pietersburg industry and other sites reveal that the two industries overlapped. In KRM Cave case, MSA I-III produced archaeological materials described as characteristically the Bambata complex industry (Singer and Wymer 1982:202).

Howieson's Poort Industry. The distinct stone industry known only one layer at Klasies River Mouth and at several other South African localities is called the Howieson's Poort Industry after the site of Howieson's Poort. Like the Bambata complex industry, the Howieson's Poort industry is characterized by small scrapers, gravers and notched pieces. However, Howieson's Poort assemblage also contain large numbers of very small blades delicately trimmed into specialized forms such as crescents, trapezes, triangles, and obliquely blunted points all of which are absent in the Bambata. The Howieson's Poort industry from the Klasies River Mouth appears in Layers 10 to 21 at Shelter 1A and Layers 1 to 5 at Cave 2. The industry is made on finer-grained non-local materials. The upper and lower layers show considerable differences in the proportions of size range and in the subtleties of manufacture. The presence of many diffused bulbs of percussion on the flakes suggests that intermediate punches of a softer material were used to produce flakes and flake-blades. The flakes of this industry show plain striking platforms rather than faceted ones. Pointed flake-blades are rare. The cores are mainly flat, discoidal types (Keller 1973:71-74; Singer and Wymer 1982:92-95). Finally, denticulates are rare and the relative frequencies of most other formal tool types vary widely (Volman 1984:207). The Howieson's Poort industry has been considered to be an expression of the Second Intermediate Period (Vogel 1970). However, research at KRM (Singer and Wymer 1982), Boomplaas Cave (Deacon 1979), Montague Cave (Keller 1973), and Border Cave (Cooke et al, 1945) reveal that the Howieson's Poor assemblages are both preceded and succeeded by MSA assemblages. As a result, it is believed that the industry is now generally considered to be a variant of MSA tradition.

Final MSA/Post-Howieson's Poort. The industry of this period was first called Smithfield A by Goodwin and van Riet Lowe (1929). It is characterized by the production of large broad flakes, large single-platform cores, and by the rare production of blades. Blades are generally broad and irregular, indicating that they are the accidental byproducts of the flaking process. Like Howieson's Poort assemblages, the frequency of unfaceted flake platforms is high. Many flakes contain the cortex on their exterior surface. Although whole flakes or fragments were marginally trimmed to form scrapers, numerous flakes reveal that they were used directly without prior marginal trimmings (Sampson 1974:259-260). Levels 4-15 of Die Kelders (Schweitzer 1970; Avery et al. 1997), above Surface 6 at Montague Cave (Keller 1973), and MSA IV of KRM (Singer and Wymer 1982) have all produced final MSA assemblages.

The Later Stone Age. South African artifact characteristics of the Later Stone Age were first defined by Goodwin and Van Riet Lowe (1929:7). According to them, the assemblages are similar to the European Mesolithic (therefore, including microlithic stone artifacts) and characterized by the use of a plain striking platform, the production of parallel flakes instead of points, and the use trimmed end scrapers (Godwin 1946:75). Based on the recent excavations, the earliest microlithic tool making tradition appeared both in Border Cave (Beaumont et al. 1987) in South Africa and Matupi Cave in Zaire (van Noten 1977) by around 40,000 years ago. Compared to the Middle Stone Age assemblages, those of the Later Stone Age assemblages are, in general, smaller in size of flakes, low production of blades and low frequency of prepared cores.

Microlithic assemblages in South Africa are recovered from many sites and dated between 40,000 and 12,000 B.P. Deacon (1984:228) classified microlithic assemblages found in South Africa as three groups based on the frequency of the recovered microlithics. According to her, microlithic assemblages between 40,000 and 12,000 B.P.

are characterized by the use single platform bladelet cores and bipolar cores. Only few unstandardized scrapers and backed microlithics are used. Bored stones, polished bone points, ostrich eggshell beads, tortoiseshell bowls are found in these assemblages. Between 12,000 B.P. and around 8,000 B.P. Flakes produced by Microlithic flaking techniques becomes rare or absent. After 8,000 B.P. fully developed microlithic techniques with numerous tools appear.

Middle and Later Stone Age in Eastern Africa. Compared to MSA research in Southern Africa, MSA study in East Africa is less well developed. Understandably most archaeological interest has centered on the Earliest Stone Age there. However, as noted above, the Middle Stone Age archaeological record in East Africa has recently attracted interest because the new evidence implies that anatomically modern humans (*Homo sapiens sapiens*) may have evolved there during the later Middle and early Upper Pleistocene. Moreover, paleoanthropological evidence indicates that these anatomically modern humans evolved not in Europe but in Africa (Robertshaw 1995:55,57). As the Middle Stone Age, which falls within this time range, it has attracted attention as the expression of the behavior of the earliest anatomically modern humans.

Final Earlier Stone Age/Transition from ESA to MSA in Eastern Africa. In most of Angola, northern Zambia, and some parts of East Africa, the stone tool assemblages of the final Earlier Stone Age period appear to be Sangoan or Sangoan-like industries. Some scholars call the industries of this period "Acheulio-Levalloisian" (Clark 1954:156-157). In Eastern Africa, the Sangoan or Sangoan-like stone tool assemblages are produced at several sites in the Lunda region in North-Eastern Angola, Lake Eyasi basin in Tanzania, and Muguruk in Kenya. Generally, the stone tool assemblages from these sites are not typically Sangoan. Some, such as those from Lunda Region, are characterized by a large number of trimmed elongated picks, thick hand axes, core-axes, and side scrapers (van Noten 1982:47). Others, including those from Lake Eyasi Basin, do not produce core-axes (Mehlman 1987:151-158). In the present study, the Sangoan industry is included in the final Earlier Stone Age. In general, the status of Sangoan material from East Africa has yet to be resolved.

Middle Stone Age Industry. The regional diversification and variation in terminology based on local industries makes the study of the Middle Stone Age in East Africa very complicated. In contrast to the various well-known MSA industries of Southern Africa such as Pietersburg, Bambata, Lupemban, the Middle Stone Age in East Africa must be lumped together under the term Middle Stone Age. Except for the so-called Lupemban Complex industry, typologically and chronologically MSA industries in East Africa still remain poorly known because the archaeological sites in these periods are distorted or mixed. In terms of artifacts, many site collections do not represent whole assemblages but are merely selections

made by the excavators. Others collections are too small to permit effective inter-site comparisons. In addition, the details of the geographical distribution of these industries remain to be fully established.

Despite these difficulties, It is believed that the Sangoan industry develops into the lower and upper Lupemban industries during the Middle Stone Age in Angola, northern Zambia, and some parts of East Africa (Clark 1969; Clark and Haynes 1970; McBrearty 1988; van Noten 1982). The tools of the Lupemban industry are smaller and better made than they were in the Sangoan. The two most common tool types are some bifacially trimmed axes with a gouge-like working end and thin-sectioned long lanceolates. The later Lupemban industry shows the Levallois technique and pressure flaking. The appearance of tanged points, increasing use of blades and decrease in flake production from prepared cores are additional characteristics of the later Lupemban. The Lupemban complex industry showed decrease in artifact size through time and evolved into Later Stone Age industry, called Tshitolian Industry (Clark 1970:135, 162, 164).

Apart from the Sangoan-Lupemban tradition, stone tool assemblages of MSA over most of East Africa are characterized by the presence of scrapers, flake points and prepared cores. A well-known site containing this type of industry is Porc Epic cave at Dire Dawa, East-Central Ethiopia (Clark and Williamson 1984:44-64). T he three major types of artifacts produced by this site include retouched points, scrapers, and flakes carrying use wear. Burins, backed flakes, and borers are also found. The Nasera Rockshelter in Tanzania produced not only a few MSA artifacts, but contained typologically very diverse small scrapers, retouched points, chisels, backed flakes, and various cores of bipolar, single platform, discoidal type. The site supplies a long sequence from MSA to LSA (Mehlman 1977:117).

Middle Stone Age Technology

Conventionally, it is believed that, in the Lower and Middle Paleolithic/MSA period, cores were intentionally prepared in order to produce predetermined flake shapes. The term "prepared core" refers to "the production of flakes that have particular desired forms or morphologies that were preconceived in the minds of the flint knapper" (Debénath and Dibble 1994:23). The manufacture of the desired blanks and the specific plans applied to the cores before the removal of the blank from the core of this period is known as the Levallois technique. For more than a century prehistorians have emphasized this reduction strategy based on the close morphological relationship between the core and one specific predetermined blank, that is, the centripetal preparation of the Levallois surface and the special preparation of the striking platform. Unfortunately, this morphological emphasis had lead to the misunderstanding of the Levallois technology (Boëda 1995:41-45; Copeland 1983); Inizan *et al.* 1992:48; van Peer 1992:1-5). As a correction, recent research on the

Levallois technology has focused on volumetric cores, which are characterized by flat upper and convex under surfaces.

Boëda (1995:46-52) proposes six technical criteria as the characteristics of the Levallois core (Figure 21). First, the volume of the core consists of two asymmetrical convex secant surfaces. Second, of the two surfaces, the flat or upper surface produces varied blanks that are predetermined, while the more convex or under surface serves as a location of the striking platforms used to produce the blanks. Third, as the maintenance of the lateral and distal convexities serve to guide the shock wave of each predetermined blank, the flaking surface is maintained. Fourth, the fracture plane of the predetermined blanks and the plane of intersection of the two surfaces are parallel to each other. Fifth, as the surface of the striking platforms must be oriented in relation to the flaking surface, the line, called hinge, created by the intersection of two surfaces is perpendicular to the flaking axis of the blanks. Based on this element, the under surface is maintained although there are various ways of maintaining the lower surface. Finally, only one technique of flaking, direct percussion with a hard hammer, is used in the Levallois operation. Figure 22 schematically illustrates the Levallois method. The manufacture of Levallois

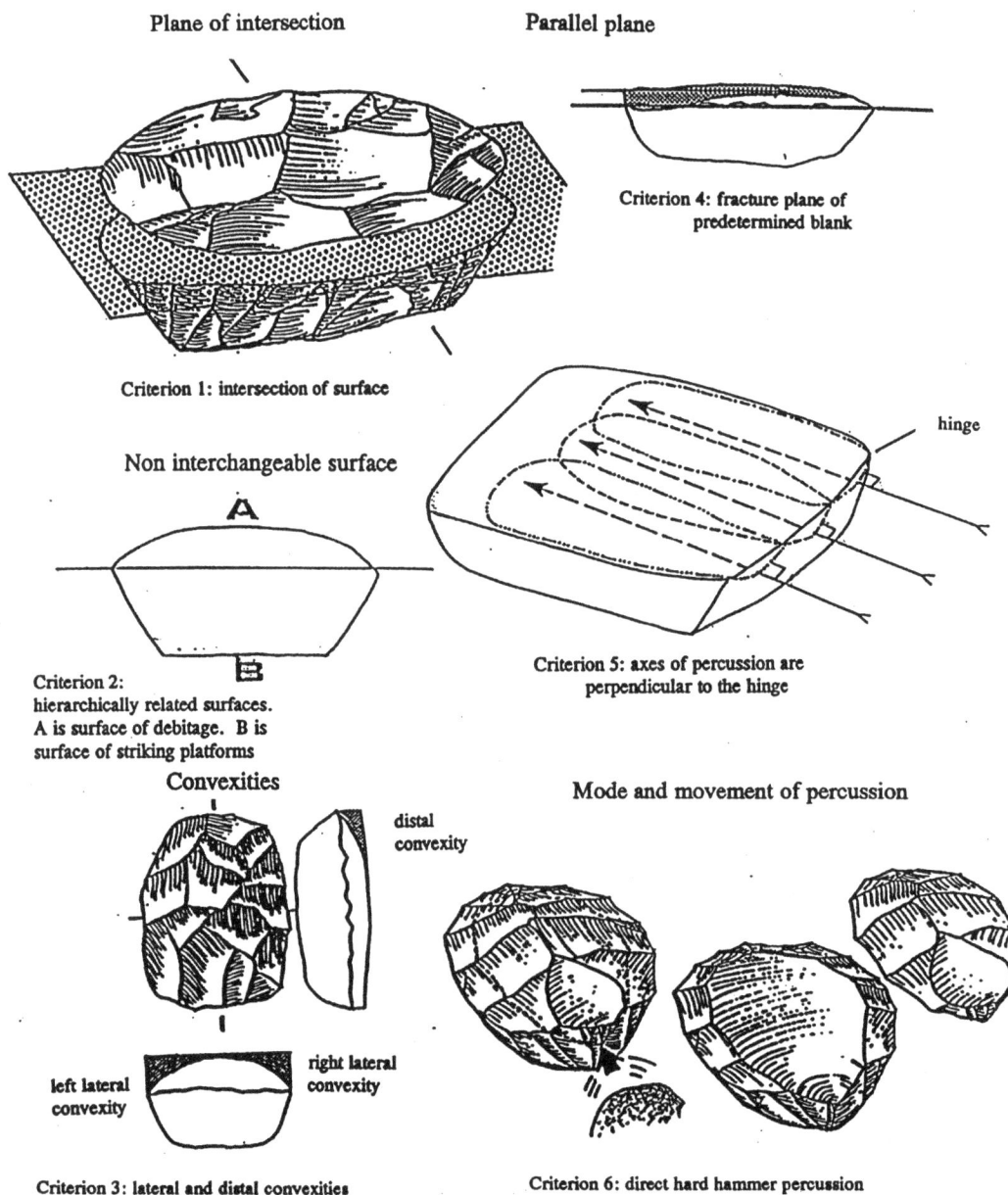

Figure 21. Criteria of Levallois Concepts.

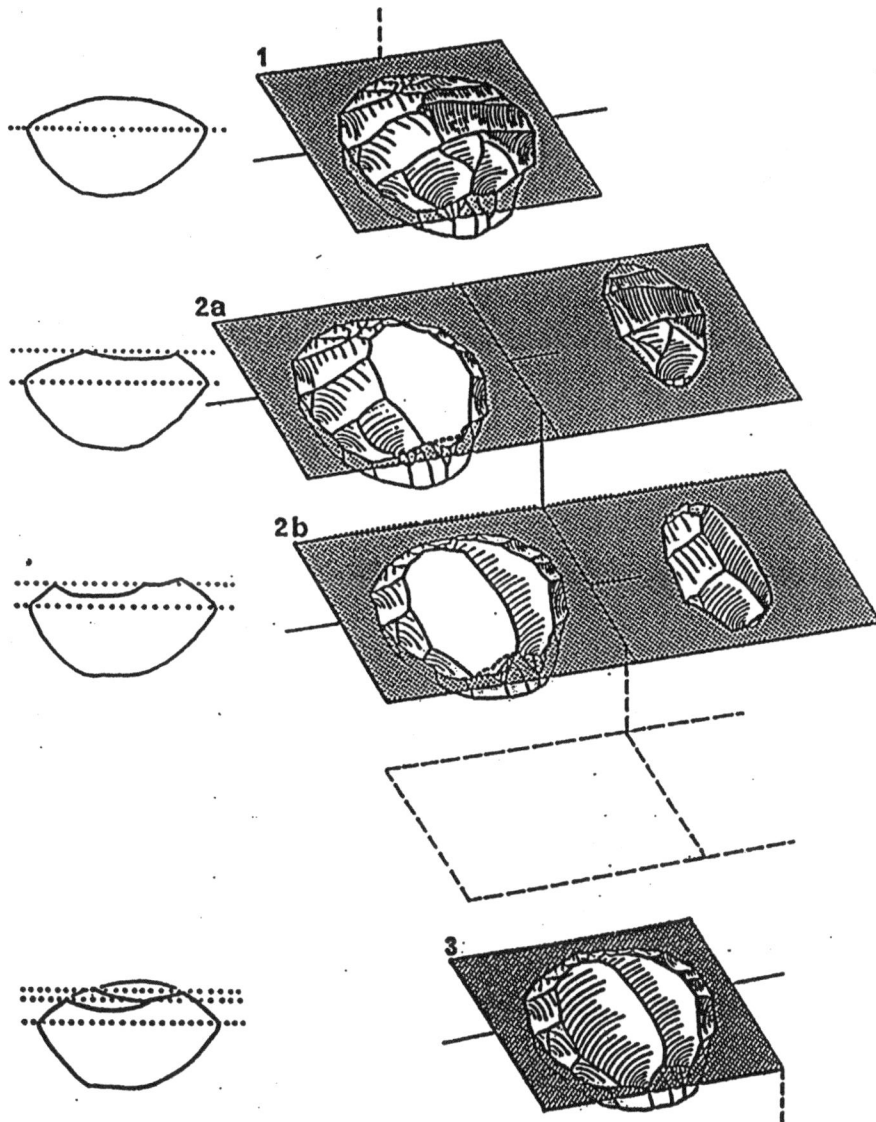

Figure 22. The Schematic Operation of Levallois Method.

blanks consists of two methods. One is the production of a single flake fromeach prepared core surface, and the other is production of a series of Levallois flakes. In the former case, the striking platform of the flake is relatively small compared to the total upper surface area. The predetermined flake is large enough to cover almost of all of the upper surface. In the latter case, the manufacture of predetermined flakes or blanks depends on the preceding removal. In this case, the multiple flakes produced in this way have different morphologies depending on (1) the orientation of their removal and (2) on the positions and size of their striking platforms. The flakes can be produced by recurrent unipolar, bipolar, and centripetal Levallois methods. Recurrent unipolar Levallois method tends to produce elongated flakes, which have a single direction of removal. The striking platform is small.

When the directions of the removals converge, the flakes are triangular in shape. The recurrent bipolar Levallois method characterized by two opposed preferential striking platforms is applied to produce Levallois blades. In recurrent centripetal Levallois method case, the entire edged of the surface can be used as the striking platform. This method is confused with discoidal core method (Copeland 1983:17-18; Inizan *et al.* 1992:53). Although the advantage of Levallois technology is still ambiguous, scholars suggest that the method allowed prehistoric people to obtain the largest blanks possible form a particular surface. Furthermore, in terms of raw material economy, the method is very efficient because people could produce a number of flakes without waste from the predetermined core (van Peer 1992:114-116).

G-Young Gang

Later Stone Age

The transition from the Middle Stone Age to the Later Stone Age in East Africa is characterized by a gradual appearance of microlithic technology is in general gradual change in East Africa. The gradual change of lithic industry is evidenced by several sites. The LSA sites include the Layers above 5 of Nasera Rockshelter in Tanzania (Mehlman 1977), Matupi Cave in Zaire (van Noten 1977), and Lukenya Hill in Kenya (Gramly 1976). Of these sites, Matupi Cave is the most important site because the lithic industry of Matupi Cave is characterized by microlithic nature in flakes, cores, and *esquillées*. The industry includes small scrapers, borers, burins, notched implement, and backed flakes or geometrical microliths and bladelets. The lowest level of the site that produced microliths is radiocarbon dated about 40,000 B.P. (van Noten 1977:35-39). As a result, the beginning of the Later Stone Age in East Africa is pushed back to 40,000 B.P. Microlithic industries in East Africa are found in various sites with various times. However, the Later Stone Age was established at least by 20,000 years ago over all of the East Africa (Robertshaw 1995:61,62).

Despite the regional diversity evident in local industries, the technological changes and development in South and East Africa, in general, parallels one another. Through time, in both regions heavy-duty large core tools decrease and finally are replaced by flakes, blades, and flake points produced from the prepared cores or Levallois cores. Finally, this flakes and prepared core technology is replaced by backed microlithic technology.

Appendix B
Methodology

In this chapter the various approaches to the analysis of the debitage, cores and tools is discussed. It begins with a review of the various approaches to debitage analysis, including individual flake analysis and mass or aggregate analysis. In core analysis, the nature of cores and core assemblages are examined. In tool analysis, functional and stylistic variability in the lithic assemblage is reviewed. Functional variability is treated in terms of edge angle and edge wear, and stylistic variability is approached from the standard, isochrestic, and iconological perspectives. In the final section, the research design used in this project is presented based on the various approaches to the lithic assemblages described in the main body of the chapter.

Debitage Analysis

A core is a piece of stone from which flake has been removed. It is a blank or unmodified piece of raw material, which includes a spall, nodule, irregular chunk cobble, or any other form suitable for the end product (Callahan 1996:36). A flake is the piece that comes off the core (Whittaker 1994:14). Any flake can either be used in its unmodified state or transformed into various types of modified pieces. All deliberately retouched or modified flakes are, in general, considered tools (Debénath and Dibble 1993:10). The process whereby chipped stone flakes are removed from lithic raw material and transformed into the useful objects is called the reduction process because stone must be reduced working until the desired form is achieved. Generally, the more complex the form, the greater the amount of reduction required (Collins 1975:16). When lithic material passes through this reduction process, the resulting pieces are classifiable as cores, tools, and flakes and a considerable amount of unclassifiable debris or "chunks." Most chunks are the result of flakes shattering in unexpected ways due either to improper reduction techniques or to physical imperfections in the raw material itself (Masse 1980:147). However, debris can also be produced from physically perfect raw material by proper reduction techniques.

Although tools and cores have received most attention from students of lithics, flakes and debris, collectively called debitage, have come into prominence recently. Debitage is a broad category that includes all flaked stone materials--both flakes and debris--which are not recognized either as cores or as tools. Debitage is produced as the waste product in core preparation, in the preliminary stages of tool manufacture, and from the occasional modification of tools during their lifetime (Fish 1979:3). It represents the material expression of a wide variety of activities and the various stages of progress of the raw material from the original form to the finished stage (Crabtree 1972:58; Fish 1981:374). Since much debitage was discarded and ignored subsequently to its creation, it is often in primary *in situ* location when found in an undisturbed archaeological context (Fish 1979:3). Finally, since debitage represents the major portion of the total artifact assemblage at most Paleolithic sites, it is generally present in sufficient quantities to support statistical treatment. Debitage can thus provides important information for reconstructing prehistoric lithic technology and patterns of human behavior.

Although debitage is a major component of lithic assemblage and an important source of information about prehistoric technology in Paleolithic sites, systematic study of debitage is rare in the African archaeological literature. However, studies done in North America, Middle East, and Western Europe have isolated certain attributes of debitage that represent a broad range of flake morphology and allow many analytical questions in archaeology to be answered by through quantitative measurement of these attributes. The most common debitage attributes include size, platform type, the presence or absence of a cortex and distal end of a flake, the angle of a striking platform, and so on. By describing details of shape, platform, and facial characteristics, which provide imprint of the discrete acts of knapping, we can understand the knapping behavior used to produce a particular tool and core. The absence of tools or cores discarded during the production process complicates the reconstruction of specific knapping activities at a site. However, debitage analysis can still tell us something of the activities. For example, fluted and notched flakes can provide direct evidence of the use of pressure technique. Debitage from the production of blades can be taken as indicating various types of tools such as scrapers, burins, and knives were being made. Levallois points also indicate that special preparation of cores was required prior to the production of flakes. Therefore, taxonomies of lithic debitage developed in the regions referred to above are useful in formulating research design of debitage analysis in Africa.

Debitage has been analyzed in various ways depending on the purpose of the lithic studies. The methods of debitage analysis can be largely divided into two classes: (1) individual flake analysis and (2) flake aggregate analysis (Ahler 1989:86). The first type includes decortication or core reduction debitage analysis, tool debitage analysis, and hierarchical analysis. The second type is called simply Amass analysis.

Individual Flake Analysis: Decortication or Core Reduction Debitage Analysis

The major premise of core reduction debitage analysis is that all debitage is produced in the core reduction process (Simmons 1983:165). The most widely used system of classifying core reduction debitage uses three stages. The first flake removed from a core generally preserves cortex on its entire exterior surface. This flake is called a primary flake and is presumed to have been produced by the direct percussion technique. Subsequent flakes removed from the same core are considered primary flakes if they do not overlap each other (Debénath and Dibble 1993:10). The removal of primary flakes constitutes the first step of initial preparation or reduction of the core. As further flakes are removed from the core, diminishing amounts of cortex are visible on their exterior surfaces. These flakes are called secondary flakes. Tertiary flakes are those removed after secondary flakes. They have no cortex on their exterior surface or only a small amount of it. The amount of cortex on tertiary flakes varies between analysts making their definition very subjective (Ackerly 1979:320; Brown 1991:371-372; Simmons 1983:165-166).

Lithic analysts use core reduction debitage analysis of the proportion of cortical debitage in assemblages to reveal differences in how particular raw materials arrived at the site. In general, they assume that low frequencies of cortical waste indicate that much of the cortex of roughed-out cores or tools were removed at some other places, presumably the source location, before the material was transported to the site. Likewise relatively low frequencies of shatter and cortical flakes are assumed to indicate that only tool manufacture, rather than primary decortication of cores, was undertaken at the site. In addition, low frequencies of cortical flakes with relatively numerous angular pieces suggest preliminary decortication of nodules elsewhere with significant core reduction occurring later at the site. Since the amount of debris or chunks reflect the extent to which core reduction contributed to the formation of an assemblage, the core reduction or non tool Debitage analysis are considered as indicating whether a given raw material was imported predominantly as unworked nodules, as roughed-out cores, or as tools which passed through a primary preform (Brown 1991:377-381). A primary preform of tool blank is characterized by a symmetrical handaxe-like outline (Callahan 1996:vii).

The premise that three stages represent an invariant sequence of flake removal has been questioned by some scholars, notably by Stafford (1980:266) and Sullivan and Rozen (1985:756). They note that the proportion of cortex that defines each debitage category is unstandardized because primary flakes may be removed at any point in a reduction sequence. Further, if a blank used as a core does not have any cortex on it, the determination of its sequence of core reduction becomes uncertain (Errett Callahan, personal communication 1997). In addition, proximal or distal flake fragments classified as tertiary flakes might actually be secondary flakes, simply missing their cortical sections.

Further, Sullivan and Rozen (1985) contend that cortical variation is only indirectly related to technology because the amount of cortex on the flake does not tell the specific technique used to produce the flake itself. They assert that variation in the proportion of cortical debitage in an assemblage depends on such factors as raw material type and availability, core size, intensity of reduction, the nature of regional raw material procurement and reduction systems and stylistic and functional factors as well as the order that the flakes were removed in the three stage reduction sequence.

Despite this criticism, the presence of cortex, the quantity

of the cortex, and the position of cortex on the exterior surface of flakes remains valuable sources information about the core reduction sequence at a site. The importance of these valuables and how these variables are integrated into the core reduction process are described following section.

The Presence and Quantity of Cortex. Cortex is particularly well developed on the surfaces of raw material in nodular form. The early reduction of nodules, in general, results in the removal of more flakes with cortex than later reduction. In fact, most lithic reduction sequences begin with decortication of cores in order to obtain non cortical portions of raw material for the manufacture of the desired form of blanks. In core reduction both the primary (cortical) and secondary (partially cortical) flakes can be assumed to have been detached at the outset of the sequence. Cortex removed later in the core reduction sequence is less frequent and is presented smaller and smaller in amount. Although there is the possibility that any type of flake may be produced in any core reduction sequence, it is nonetheless true that the probability of removing cortical flakes in the early stage is high and in the late stage is low.

The Location of Cortex. The position of cortex on the exterior flake surface may indicate how the cortex was removed from the nodule. When one prepares different types of cores, the removal of cortex may proceed by quite different methods. These methods can be reflected by the position of the cortex on the exterior surface of the partially cortical flakes. For example, the use of one or more platforms on a core at the time cortex was removed may be reflected in the position of cortex relative to the platform of the flakes. Figure 23 compares the removal of cortex from a single platform with multiple platforms. As Figure 23 illustrates, if cortex was removed from several core platforms, there is a higher incidence of flakes with cortex covering the proximal half of the flake but lacking cortex on their distal ends (Baumler 1987:54-56). Therefore, the location of cortex on the flake indicates the number of platforms on a core to produce flakes.

Individual Flake Analysis: Tool Debitage Analysis

Several key morphological variables including: shape (Masse 1980:146), platform characteristics, relative thickness, size and shape, platform angle, flake scar pattern, and retouch type (Simmons 1982:35) are also usually analyzed. The analysis of these morphological variables is called tool debitage analysis.

Flake Platform Analysis. The flake platform is the remnant of the original striking platform of the core from which the flake was removed. The type of the platform on a flake may therefore provide information about the stage of manufacture at which the flake was removed. Unfortunately, the platform is often obliterated beyond recognition by crushing when it is removed; if the flake has snapped on removal, it may be missing altogether. On flakes that retain their platforms, at least five platform types are recognizable: (1) platforms composed of cortex, (2) platforms consisting of a single facet, (3) platforms with more than one facet, (4) platforms with retouch scars which are perpendicular to platform width and originated prior to the flake's removal from the core, and (5) platforms retouched after the flake is removed from the core (William Dickinson, personal communication 1997).

When a flake is removed from the parent material that still retains its cortex, the striking platform of the flake will exhibit cortex. If the core has been prepared by removing a portion of the parent material to produce a relatively flat surface for use as a striking platform, the resulting platforms will not exhibit any cortex. Flakes removed from such parent material will have a platform consisting of a single facet or flat surface lacking cortex. If some flakes have been removed from the perimeters of large blanks which have been previously retouched to produce useful flakes, the platform on the removed flake will exhibit a tiny portion of the retouched edge on the blank or the larger piece of the flake. Therefore, the type of flake platform can tell us whether a flake was removed from unprepared parent material or prepared and retouched cores. It also provides us with information as to the extent to which cores or blanks were retouched to produce flakes (Chapman 1977:371).

Relative Thickness. In basic flint knapping, it is often assumed that the hard hammer technique produces thicker flakes than those produced by the soft hammer technique. Huckell (1973:123) states that (1) flakes that exhibit thin and strongly lipped striking platforms indicate that bifacial thinning was done with a soft hammer and (2) larger and proportionately thicker flakes with larger, unprepared, unlipped striking platforms and prominent bulbs of percussion indicate the use of a hard hammer. However, analyzing flakes in this way is too simplistic as these attributes depend on other factors, such as morphology or dorsal surface, as well as the type of percussor used in their production. For instance, a soft hammer can also produce flakes with prominent bulbs of percussion and strong lips. If a striking platform is located just above the ridge on a core, a soft hammer blow produces flakes with strong bulb of percussion. If a striking platform is located between the ridges on a core, a soft hammer produces flakes with strong lips (Errett Callahan, personal communication 1997). Relative thickness of flakes may therefore indicate the stage of reduction sequence rather than the type of precursors used to produce them.

Flake Size and Shape (Platform). Flakes produced as a byproduct of knapping behavior cannot be larger than the maximum dimension of the parent material or spall. A spall is a relatively large flake blank to be reduced down into a biface core (Callahan 1996:viii). As reduction continues during the manufacture, use, and maintenance of a stone tool, the tool becomes progressively smaller. During this process both the maximum possible size and

average size of the flake byproducts decrease progressively as well.

In general, both hard and soft hammer percussion techniques produce flakes that are much larger in size than those produced by the pressure technique (Ahler 1989:87). Thus, flake sizes can provide us with the information about the stages of reduction process and the techniques that were applied to the removal of them. However, one must be cautious; there are exceptions to these rules as well. The length of flake largely depends on the knapper's skill as well as the size of end product. Experiment shows that large and small flakes can be produced by both percussion and pressure techniques. In fact, percussion techniques can actually produce smaller flakes than those produced by pressure, and pressure techniques can produce larger flakes than those produced by percussion technique. Practically speaking, however, we know that, throughout reduction sequence, percussion techniques are more widely used than pressure technique in initial decortication and the early reduction stages, while pressure technique is preferred to percussion technique in the late reduction stage. In addition, as the end product is smaller than the initial tool blanks or cores, flakes produced in late reduction stage invariably become smaller and smaller.

The overall size and shape of the removed flakes also provide information about core reduction sequence because they are directly affected by, and then affect, changes in core morphology. In fact, each flake contains information of minimum dimensions of the core face at the time it was removed. For instance, blank length may approximate the maximum size of the core in at least one dimension. As the core is reduced these dimensional parameters are changed and then affect the potential dimensions for the subsequent blank. As a result, the production of large flakes becomes much more restricted as core size reduces through the reduction sequence, although small flakes may have been produced at any reduction sequence.

As flake size and shape are imprinted on the core after their removal, they can be used to reconstruct the core reduction sequence. In general, the nature of scar on preparatory flakes is useful in predicting the type of blank that requires core preparation for its production. For instance, the major key to the production of prismatic blades and their subsequent detachment is the removal of long, parallel-sided preparatory flakes. On the other hand, the removal of a flake blank can also suggest the necessity for and type of preparatory flakes that will follow its detachment (Baumler 1987:57-58).

Platform Angle. Platform angle is also an important attribute because it can affect the size and shape of flakes. In Dibble and Whittaker's experiments (1981), flake termination, length, and thickness were strongly influenced by platform angles set between 30 and 90 degrees. Platform angle consists of interior and exterior platform angles (Figure 24). Interior platform angle is the angle

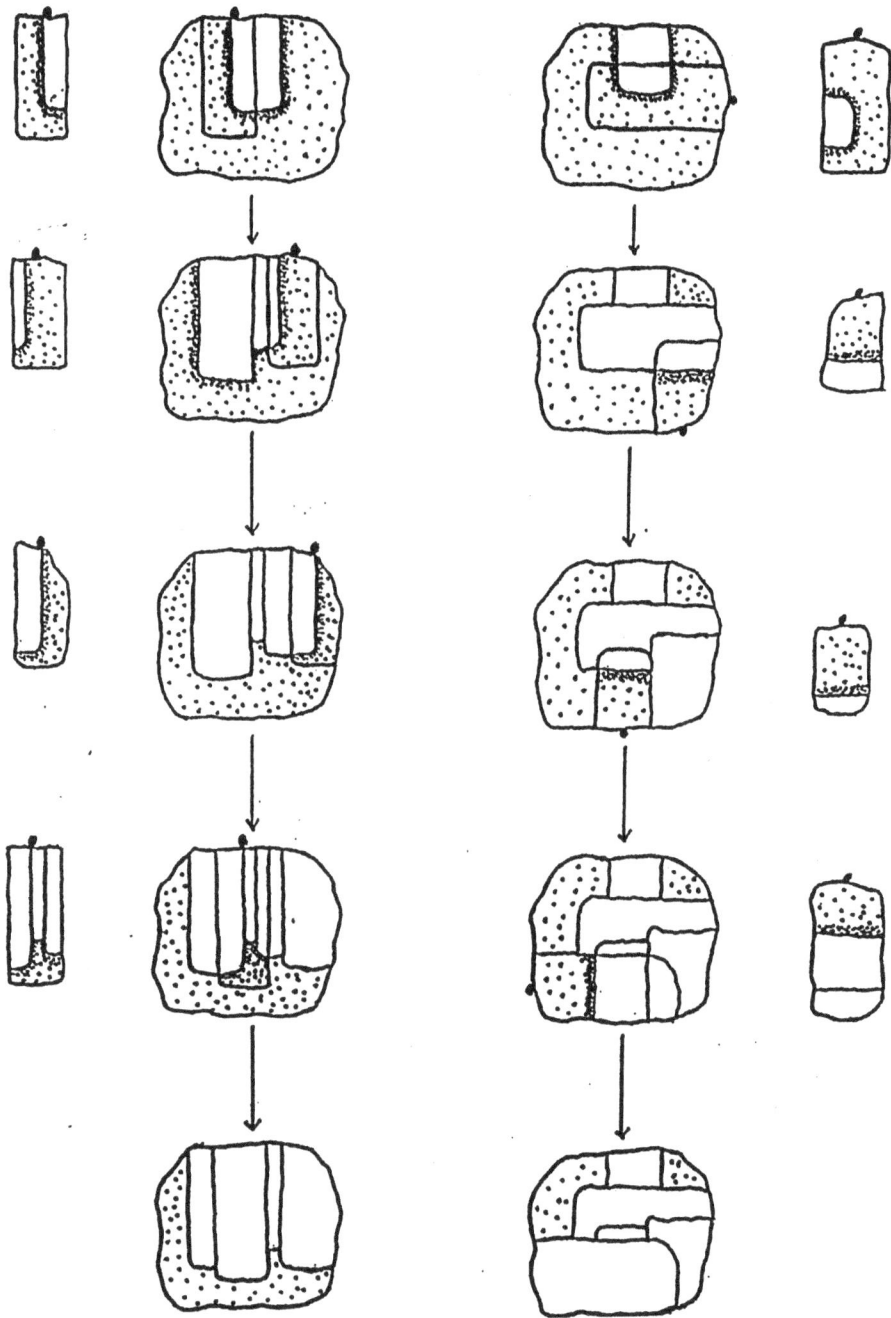

Figure 23. Graphic Illustration for the Flake Removal of Cortex from a Nodule by Single and Multiple Core Platforms.

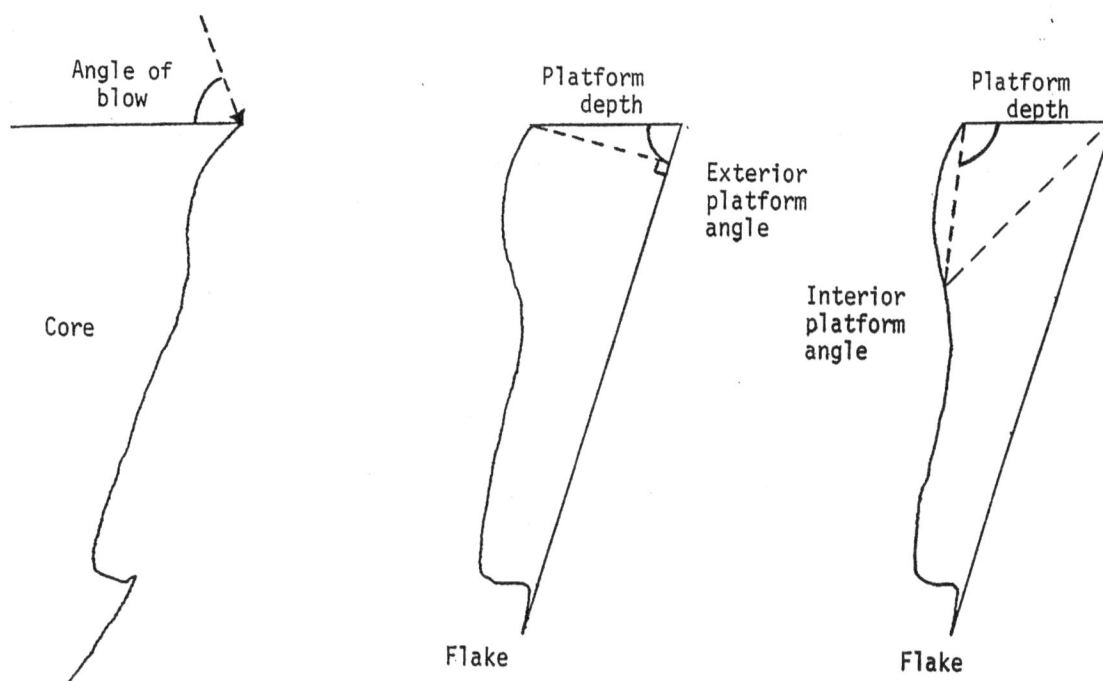

Figure 24. Exterior and Interior Platform Angles.

formed between the platform surface and a line through the point of percussion to the base of the bulb. Exterior platform angle is the angle formed by the intersection of the platform surface and the exterior surface of the flake. Of these two platform angles, exterior platform angles largely affect flake morphology. Here "angle of the striking platform" means only exterior platform angle. In the case of flake termination, feathered terminations occur at lower platform angles than overshot flakes. For the lowest platform angles, the interior and exterior surfaces of the flake tend to converge and the flake terminates at the intersection of these two surfaces in regular fashion. For the middle range of the platform angle, two surfaces are essentially parallel. For the highest platform angle, the interior and exterior surfaces of the flake diverge. That is, in the latter two instances the flake tends to terminate in a hinge fracture or an overshot. At the highest platform angles, complete flakes are rarely produced because flakes can easily overshoot or fail to develop completely. A more detailed explanation about fracture mechanics and definition of flake termination type will be provided in the next chapter.

Platform angles can also affect flake length and thickness. When impact force is constant, flake size depends largely on the combined effects of platform angle and platform thickness. If platform thickness is constant, an increase in the platform angle results in a longer and thicker flakes. When the platform angles are low to moderate, an increase in platform thickness increases flake size (Speth 1972). However, as the platform angle increases, the variability of flake size for a given platform thickness increases.

Conversely, as platform angle increases, there is a decrease in platform thickness that will permit the detachment of a flake. In other words, an increase in the platform angle with a concomitant reduction in platform thickness to a relatively small value results in the production of maximum flake length (Table 33). This is the strategy employed to make blades by indirect or punch percussion, wherein platform thickness can be controlled with much more precision. This strategy also applies to other techniques.

Flake Scar Pattern. Exterior flake scars indicate removals from a core made prior to the removal of the flake itself. Most of these scars are generally formed during reduction coincidental with the flake removal from the core but some of these may be formed a considerably earlier time. The orientation of these flake scars to the axis of striking platform indicates in what direction flakes were removed prior to the moment of a flake's detachment. The orientation of flake scars reflects the orientation and number of platforms on the core, and consequently the method of reduction at the time the flake was removed. On the basis of the directions of the flake scars, one can determine whether a core was being reduced from one platform or several platforms and whether these platforms were related to each other (Baumler 1987:49).

If they are large enough to be transformed into a tool, flakes removed from a core pass through the entire reduction process. In the course of this reduction, the flake scar pattern is also changed. In general, preparatory flakes detached from a core show irregular or random scar pattern

Table 33. Relationship between Platform Angles and Flake Length

Platform angle	Platform thickness	Flake length
Increase	Constant	Longer and thicker flake
Low to moderate (30 to 60 degrees)	Increase	Longer flake
High (close to 90 degrees)	Low	Maximum flake length

on their dorsal surface. Flakes which pass through many reduction sequences show more controlled uniformal pattern. For instance, in the case of bifacial thinning, the preparatory flakes from a blank show very irregular scar pattern. As this blank is transformed into well controlled biface through many reduction sequence, the scar pattern on flakes which removed from the biface begin to show certain type of scar pattern such as radial, parallel, or diagonal. Therefore, flake scar pattern is as reliable an indicator of reduction sequence (Errett Callahan, personal communication 1997).

The Number of Flake Scars. The number of exterior flake scars also provides us with a clue as to when in the reduction sequence a flake was removed. An early flake removal from the core will generally have fewer previous flake scars because its core received fewer preceding detachments. On the other hand, in the late stage when cores have become smaller and smaller, flakes show more complex scar patterns and higher scar counts than those from the early stages (Munday 1976:120).

Individual Flake Analysis: Hierarchical Analysis

Hierarchical analysis has been systematically developed by Sullivan and Rozen (1985:758-760). This method begins by categorizing flakes in accord with three dimensions: (1) the single interior surface, (2) the point of applied force, and (3) the flake margin (Figure 25). The single interior surface is identified by the presence of positive percussion features such as ripple marks, force lines, or a bulb of percussion. On debitage with intact striking platforms, the point of applied force is located where the bulb of percussion intersects the striking platform. If only a fragmentary striking platform remains, the point of applied force is indicated by the origin of force line radiation. A missing striking platform means the point of applied force is absent. Flake margins are classed either as intact or not intact. Debitage margins are considered to be intact, If the distal end of a flake exhibits a hinge or a feather termination and later breaks or snaps do not interfere with accurate width measurements. If these conditions are not met, debitage margins are not intact. Based on these three dimensions, four types of debitage are defined: (1) complete flakes, (2) broken flakes, (3) flake fragments, and (4) debris. Complete flakes are characterized by the

presence of single interior surface, point of applied force, and margins. Broken flakes show single interior surfaces and points of applied force have margins that are not intact. Flake fragments exhibit only single interior surfaces. Debris is characterized by the absence of all three dimensions. Sullivan and Rozen (1985:759) claim their typology is "interpretation-free" because the hierarchical typology consists of categories "that do not require the observer to infer the technological origins of artifacts in order to classify them.

Sullivan and Rozen (1985) emphasize two important aspects of their approach. First, it assumes that the significance or relevance of particular variables is not equivalent in all research situations. Second, because their categories accommodate the full range of formal variation in debitage and do not imply technological conclusions, inferences advanced on the basis of differences in debitage category proportions can be tested from the complete set of conventional technological variables, including flake size, platform characteristics, and cortical variation.

However, there are also drawbacks to hierarchical analysis. For example, during the present lithic analysis, I found some complete flakes that contained neither single interior surface nor point of applied force. Although these flakes were complete, following strict hierarchical taxonomy, they would be categorized as debris. It would appear that flakes produced by soft hammer technique that lack a distinct bulb of percussion on the interior surface cannot be perfectly categorized within the hierarchical taxonomy. This is because the bulb is disseminated and an eraillure scar and ripple marks are particularly subdued when a club-like rod of material such as wood, antler, and horn is used to produce the flake. An eraillure scar is a parasitic flake scar left after a small leaf-shaped flake pops off the bulb of percussion when the flake is struck (Whittaker 1994:15). Flakes produced by bi-polar technique, that is, by resting a core on an anvil and striking the core with a precursor, also exhibit these characteristics. Although theoretically the bipolar fracture forms a bulb at each end of the material, in practice bulbs of force are often not present on both ends of bipolar flakes. As a further complication, the bipolar technique often causes the bulb of percussion to shatter or be severed (Crabtree 1972:38, 40, 42). If a bulb of percussion is present at only one end

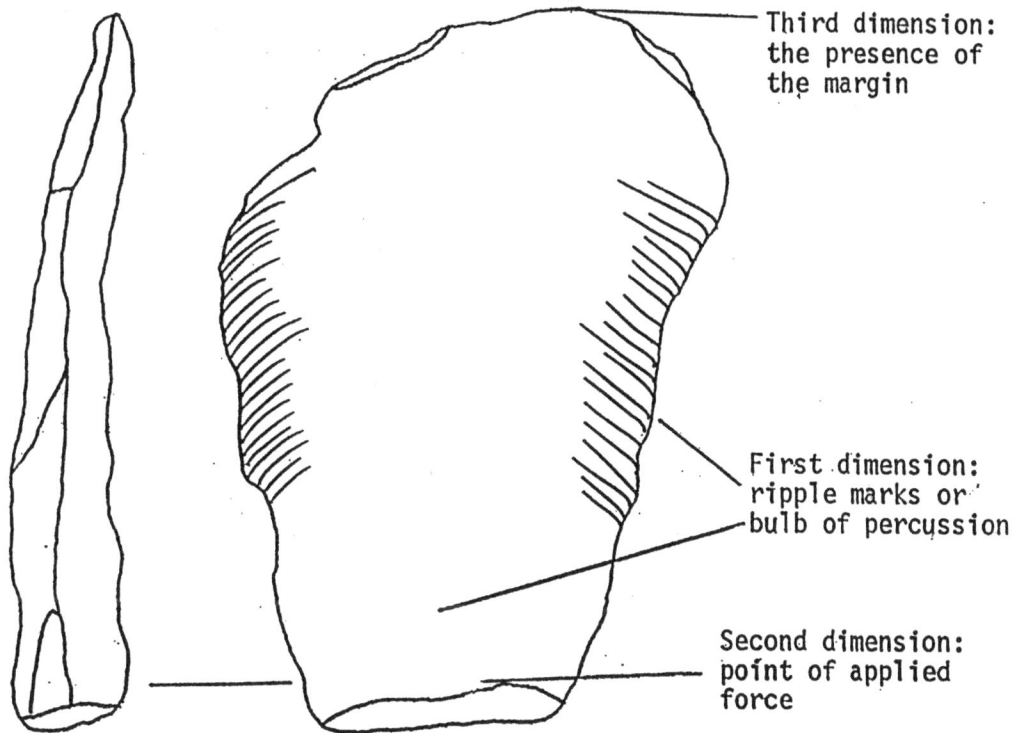

Figure 25. Three Dimensions.

of a flake, that is, proximal part, and the distal end is sheared, it is difficult to detect complete flake termination. As a result, the hierarchical analysis is also criticized.

Raw material variation can also complicate hierarchical analysis. Much of the debitage in the present study consists of quartzite and very patinated basalt. On flakes made of either type of material, the point of applied force is not always visible. This makes recognition of the second hierarchical dimension very difficult. As a result, the proportion of an assemblage classed as flakes or debris can be inaccurate.

Flake Aggregate or Mass Analysis

From a practical standpoint, individual flake analysis is disadvantageous in handling numerically large lithic samples because such flake-by-flake analysis takes an enormous amount of time and money (Ahler 1989:85). Each individual flake in hierarchical analysis is categorized into flakes, tools, cores, and debris. The categorized flakes are subdivided into complete, broken, and fragment. Further, the measurement of one flake is a complicated process. If broken flakes are also measured, the time and cost doubles. In contrast, mass or aggregate analysis does not involve classification of individual flakes or flake attributes but is based on the size-graded aggregates of flaking debris. It therefore advantages in handling numerically large samples of flaking debitage such as those collected in large-scale archaeological projects using screened artifact recovery.

The basic procedure of aggregate analysis is (1) to size-grade the complete excavated flake assemblage using a series of nested screens and (2) to collect simple quantitative data from the subsample of debris caught in each screen. In mass analysis, individual flake observations are minimized in favor of simple variables such as count and total weight of the subsample in each size grade. Using aggregate analysis has some real advantages. First, the full range of aggregate of debitage from a context can be obtained. Second, extremely large artifact samples with many small flakes can be rapidly analyzed. Third, technological bias based on size can be minimized because a broad sample of flakes produced by all techniques can be included in analysis. Fourth, a relatively high level of objectivity and replicability can be expected during the data collection process.

In addition, in contrast to the hierarchical analysis, aggregate analysis is useful to analyze flakes produced by bipolar technique. The aggregate analysis does not categories the artifact on the basis of three dimensions of the hierarchical analysis, but categories them using a series of nested screens, which catches flakes produced by any techniques.

However, aggregate analysis also has some drawbacks. A major problem with aggregate analysis is evident in the analysis of mixed samples. While individual flake analysis allows direct detection of some of technologically discrete knapping episodes, in aggregate analysis the behavioral byproducts are studied as a unit. As a result, a detailed technological attributes of the individual flake are ignored.

Using mass analysis for the present flake assemblage was rejected for four reasons. First, it can lose information about provenience of each stone artifact. The number of Middle Stone Age sites in East Africa is limited; it is therefore necessary to extract as much information as possible from each flake. To this end, during excavation of our two rockshelter sites we plotted most MSA flakes in three dimension to keep their provenience during excavation. Mass analysis would sacrifice this provenience data. Second, mass analysis focuses on size and weight, and disregards other attributes such as flake scar pattern and number of flake scar which are very useful to understanding the core reduction sequence. Third, mass analysis is generally based on the assumption of a three-step reduction sequence. That is, through the reduction sequence, size, thickness, and weight are supposed to decrease. It also assumes that percussion technique produces larger flakes than those produced by pressure, and bipolar flakes are much longer than bifacial thinning flakes produced by soft hammer percussion. But, we have seen really "rules of thumb" for which there are many exceptions. Finally, the majority of our flakes are made of basalt, which requires careful examination because the flakes are severely patinated. Under these circumstances, mass analysis is not useful for our assemblage. Therefore, individual flake analysis including decortication debitage, tool debitage, and hierarchical debitage analyses was used in the present lithic study.

However, it is almost impossible to determine the standard nature of the lithic debitage at any given site from literature dealing with the Middle Stone Age in East Africa. That is, because of the absence of basic date to compare, it is difficult to do a comparative study involving debitage. In addition, a comprehensive body of information about the African Middle Stone Age does not exist which incorporates sufficient uniformity to allow comparisons from site to site. One of the important goals of lithic debitage analysis in the present study is to provide quantified description on which to base further investigation. Such quantitative description or analysis of debitage attributes of single site or stratum within a site will provide information about a number of technologies of local cultural behavior.

Core Analysis

A core is a nucleus or a mass of material preformed by the knapper to produce the desired shape of flakes or blades. Core scars reflect the detachment of one or more flakes. Cores, like flakes, indicate the process of flaked tool industries (Crabtree 1972:54) and the transformation of a core into desired or useful shape of flakes is the reductive

technology (Collins 1975:21). This technology involves the removal of cortex on the raw material through each stages until the desired form is achieved. The original mass of raw material is reduced producing an end product smaller than its initial mass. The major goal of core analysis in the present study is to reconstruct this core reduction sequence.

The initiation of core reduction follows basic fracture mechanics, which depend on the laws of physics and the properties of conchoidal fracture in an elastic but brittle and homogeneous mass. On the basis of controlled experiments and replications Crabtree (1972), Cotterell and Kamminga (1979), Dibble and Whittaker (1981), Speth (1972, 1974), and Whittaker (1994) have isolated the variables that affect the production of flakes and the detachment of flakes from cores or blanks. The basic principles of fracture mechanics are described in the next chapter. This is because the fracture mechanics is related not only to core and blank preparation, but also to flake production and break patterns caused by use-wear.

The Nature of Cores

The transformation of raw material spalls into cores begins with the detachment of the first flake. In the early stage, cores are generally large and heavy as they developed with few scars on the major removal face and simple scar patterns. Experiment shows that at this stage cores tend to produce flakes of irregular shape and size and that the ratio of large to small flake production is higher than that of the late stage. In the middle stage, both core size and removal face area on cores is reduced. The removal face is more complex both in terms of scar count and pattern. This is because, as they pass through each stage, cores come to carry the scars produced in previous stages. In the late stage, cores become too small to produce flakes. Cores in this stage are considered exhausted. The term "exhausted" is synonymous with "used up" or "consumed." Cores in the late stage of reduction often have an amorphous shape and are characterized by very complex scar patterns and high scar counts. Therefore, the number of potential tool blanks or large flakes that can be produced from a core is finite (Baumler 1987:27; Munday 1976:120).

Variables for Reconstructing Core Reduction Sequences

In general, studies of the core reduction process or sequence focus on the subsequent modification of flake blanks into various tool forms (Bradley 1975; Collins 1975) and on specific core industries such as blade cores (Newcomber 1975) and Levallois cores (Van Peer 1992). However, non-blade and non-Levallois lithic assemblage yielded cores as well. For an objective reconstruction of core reduction sequence from archaeological lithic assemblages not only the core itself but also other variables such as flake scar pattern and flake size are important. This is because cores are parent materials of all tool blanks, flakes, and shatter. Baumler (1987) provides a clear description of the importance of these variables and how they are used in reconstructing core reduction sequences. His variables can be recognized in both core and flake assemblages. Core assemblage includes exhausted and unexhausted cores. Flake assemblage focuses on exterior flake surface. The following descriptions are summary of Baumler's variables (1987:45-69).

Core Assemblage. Most common core types found in archaeological sites are exhausted and unexhausted ones. Exhausted cores are, in general, discarded at the finish of each core reduction event. Thus, exhausted cores represent the end of the intended reduction sequence or sequences and are byproducts and waste. Unexhausted cores may represent various stages in reduction sequence or sequences. These cores were discarded before exhaustion which might be caused by not only core size but also steps and hinges, reduction of platform size or angle and lack of material (Crabtree 1972:62).

By using core assemblage from the archaeological site, one can determine the existence of one or more types of reduction sequences at the site rather than the course of reduction itself as cores represent, to some extent, successfully completed reduction events. For instance, homogeneity in the core assemblage, expressed in overall core morphology and flake scar pattern, generally reflects a single reduction sequence. On the other hand, heterogeneity is generally created not only by multiple sequences but also by multiple strategies in a single sequence. In the latter case, varying demands for the products a single sequence can lead to the discard of different types of cores. If one wants to produce a particular blank type, reduction may stop at various points along a single reduction sequence. As a result, heterogeneous assemblage created by multiple strategies in a single sequence will be similar to that produced by multiple reduction sequence.

In general, multiple strategies in a single reduction is often implied when the assemblage consists of two well-defined core types in which one is generally larger than the other, relatively less abundant and rare in comparison to the type of blank it appears to have produced. On the other hand, if the assemblage consists of the various core types that are similar in size and in close proportion to the blanks they produce, multiple reduction sequences can be a plausible working hypothesis. Therefore, an understanding the heterogeneous core assemblage requires a careful consideration of the core morphologies, their relative number and size, and their quantitative relationship to the blank assemblage.

However, except for the potential number of sequences described above, the cores themselves provide only a limited amount of information about the reconstructing reduction sequence itself because most cores are discarded at the time when the reduction process is completed except for the prematurely discarded cores. As a result, the information obtained from cores is an insight into the last

stages of the reduction sequence as this is reflected in the flake scars of the final preparation or last blank produced. When we consider the last stages of the reduction sequence, the orientation of core platforms and its shape and size are important variables. Particularly useful is the core size in determining which flakes are likely to have been generated early or late in the sequence on the basis of the size of the blanks in relation to that of the discarded core. However, as mentioned above, cores can tell us about the number of core reduction sequences represented and the end points of each sequence. They seldom provide us with much information about the reduction sequence itself. Therefore, flake assemblage that represents the major part of the sequence is also important in understanding the core reduction sequence.

Flake Assemblage. The single most important variable of a flake useful for the order and conditions under which it was detached from the core is its exterior surface as a flake's exterior surface is a record of the core at the time that the flake is removed. Therefore, the exterior surfaces of flakes removed at different times can provide us with information about the appearance of the core at various points in the reduction sequence.

Particularly helpful attributes of the exterior surface of flakes are the nature of the flake scars and the remnants of cortex on exterior surfaces of the original raw material. The former can be expressed as the number and orientation of flake scars. The latter as its presence or absence, quantity, and position. All these variables mentioned above are described in the section of debitage analysis.

Tool Analysis

According to Winner (1977:75), an apparatus is "a structure of material parts assembled to produce determinate, predictable results when placed in operation." A technique is "a structure of human behavior in order to accomplish an outcome." A technical organization is "an assemblage of human beings and apparatus in structured relationships designed to produce certain specified results." I am seeking to understand the technical organization of the Middle Stone Age. Of course, only one aspect of that technical organization remains: its tools. Therefore, I must reconstruct the technical organization of the Middle Stone Age on the basis of tools.

A tool is the finished form of stone artifacts or artificially retouched ones. Tools, in the chipped-stone assemblage from the archaeological sites are very important because they embody two distinct properties: those determined by the task for which it was produced and those that reflect choices made among several equally valid and feasible options. The former property is, in general, called the functional attribute (s) and the latter the stylistic attribute(s). The functional attributes of a tool are determined by such factors as how the artifact was produced, the various ways it was used, the meaning it had to its owners, and the manner in which it indirectly reflects other activities belonging to the ongoing cultural situation in which it serve (Sackett 1973:320). The definition of style varies according to many archaeologists. According to Sackett, style is:

> an aspect of every pattern visible in any archaeological situation. It is based upon the notion that there are usually alternative means of achieving the same end, that the specific expression any give artifact assumes results in a sense from a choice made among several equally valid and feasible options, and that the choice made in any given cultural situation is determined by its historico-genetic setting (1973:321).

In any study that attempts (1) to treat artifacts as a part of a functioning cultural system and (2) to define the particular traditions of two or more prehistoric cultures in a meaningful way beyond spatial-temporal distribution, a primary goal must be to distinguish between these two kinds of properties. When one compares assemblages from two or more prehistoric cultures which exploited different aspects of similar environment, one would expect that the attributes of their artifacts might differ with respect to both function and style. However, some functional attributes between cultures with similar patterns of exploitation of similar resources would be little different. Cultures closely adjacent in time and space would show small differences in both function and style. That is, if the full inventory of recovered artifacts from two different cultures show very little differences in functional and stylistic attributes, we can postulate that two cultures were closely adjacent in time and space. Therefore, one will be able to re-create systems of exploitation and even distinguish a certain degree of communication between two or more cultures on the basis of the full recovery of artifacts. However, it is hard to validate hypothesis about function and style if the cultures of the study areas produced artifacts largely confined to chipped stone tools and the debris resulting from their manufacture (Jelinek 1976:19-21).

According to Jelinek (1976), one major problem in the analysis and interpretation of materials of purely lithic sites lies in the nature of the collections themselves. There are three different kinds of sites recognizable with respect to tool manufacture and use: those primarily devoted to manufacture, those showing only selected products of manufacture, and those including both evidence of manufacture and evidence of the use of tool. Manufacturing sites are characterized by quantities of exhausted and unsuitable partially worked cores, broken or misshapen flakes and preforms, and great quantities of debris resulting from the reduction of cores and preforms. On the other hand, tool use sites yield primarily exhausted or broken tools and larger flakes with little or no debris resulted from manufacture. As a result, if the sites are characterized by the presence of exhausted and broken tools along with strong evidence of lithic manufacture,

they are generally considered indicating a wider range of activities rather than a simple manufacturing station. Despite this general sense, most lithic materials recovered by archaeologists are objects that were no longer wanted by the inhabitants of the site. The composition of lithic collection is an important in the interpreting the function and style of a lithic industry. Such composition is affected by several variables. These variables include the proximity of lithic resources, the size and structure of the population of the group involved and the length and or periodicity of their occupations. Therefore, directly assigning either functional or stylistic significance to collections is very difficult and should be approached with caution, even suspicion.

The attributes of the artifacts themselves are major variables pertinent to a study of function and style. These attributes include the raw materials from which the artifacts are manufactured, the techniques that were applied to making chipped stone artifact, and the classification of tool retouch. The raw material suitable for flaking, and its relationship with the technology will be described in the next chapter. The most important information about the technique can be obtained from debitage. The detailed description is in the debitage section.

Functional Analysis

Functional variation in the stone artifacts is the product the post-detachment modification of flakes to enhance their efficiency in the performance of certain work and of further alteration of the utilized edges or surfaces of the artifacts due to the stress of use (Wilmsen 1967:139). That is, functional variations in tools include both their initial unmodified form and the shape they come to assume through use or retouch. Much of this modification or alteration appears on the margins of the tool as a result of breakage by force to which the edges were submitted. Therefore, it is not surprising that the morphological aspects of chipped stone tools which have been studied most closely for direct functional relationships are edge angle and edge wear/use wear. Functional variations that occurred in the prehistoric time are commonly reflected in the correlation between edge angle and edge wear.

The type of work that affected edge angle and caused edge damage on the stone artifacts is also useful in understanding functional variation because different tasks of work require different edge angles and create different scar types and patterns on the edge. The type of work reflected most frequently on archaeological materials involve scraping, called pulling motions, where a tool is moved across its working edge, usually toward the operator. The angle at which the tool is held in the operations is called the "rake angle." Shaving is a movement toward the operator with short strokes transverse to the working edge. In shaving, the rake angle of tool to worked material is generally considerably less than in scraping. Whittling is similar to shaving but the direction of motion transverse to the working edge and generally away from the operator. Cutting/sawing and slicing/carving are motions longitudinal to the cutting edge. Cutting is operated unidirectionally, while sawing is done bidirectionally. The angle of slicing and carving with which the tool makes contact with the worked material is usually acute and highly variable. Two things happen as a result of this variability: differential pressure on either surface takes place and more damage to one of the faces than to the other occurs. Boring or drilling is a circular motion producing both rotary pressure to the sides of the drill and a direct pressure downward into the worked material. Chopping, including wedging and adzing, is an action in which the force is directed straight through a tool margin into the main body of the implement (Odell 1981:199-206; Odell and Odell-Vereecken 1980:98-101; Tringham *et al.* 1974:181).

Edge Angle. Different angle size ranges would be related to both to functional effectiveness and functional differences. Edge angle ranges can be grouped into types that are indicative of function. In general, cutting actions are associated acute bit ranging 26 ° to 35 °. Many tool with this edge angle values might serve as meat and skin-cutting knives. A broadly useful attribute appropriate to a number of functional applications may be related to the edge angle values ranging between 46 ° and 55 °. Inferred functional variations for this range of edge angle values include skinning and hide scraping, sinew and plant-fiber shredding, heavy cutting for bone or horn, and tool back-blunting. Both large, unhafted tools which show retouching either on the distal edge or on one or both lateral edges and socketed end scrapers are considered appropriate implements for the first set of tasks. The same unhafted tools and tools with retouch on both lateral edge may be used for second set of the task. Tools with natural edge angles of above 50 ° and edges carefully retouched to this size are likely the implements favored for bone cutting. However, when tools within this category were used for cutting, their utilized edges are characterized by chips that do not extend over either face of the tool but which appear to have been nibbled or nicked off. If the retouch on the edge is extended over either face of the tool, it would be not used for cutting but for blunting the back edge of the tool (Gould, Koster, and Sontz 1971:Table 1, Table 2; Wilmsen 1968:156-158). Tools with edge angles of 66 ° to 75 ° are considered to have used in wood working, bone-working, skin softening, and heavy shredding. One of the characteristics of tools that are used in these activities is occurrence of tips and concavities. Edge angles of less than 20 ° are seldom used because they were too delicate to sustain any significant pressure (Hayden 1979a:124-125; Wilmsen 1967:78-88, 139-149). In sum, acute angle of 26 ° to 35 ° are generally found on cutting tools, medium angles ranging between 46° and 55° on general purpose tools, and angles of 66 ° to 75 ° on heavy duty tools.

Edge Wear/Use Wear. Edge wear can be caused by any kinds of force, human or natural. However, in the present study, the term "edge wear" is confined to wear produced by human action. Use wear is thus the edge damage on stone tools that results from usage (Tringham *et al.* 1974:174). As lithics edge damage caused by nature is also an important attribute, edge damage caused by natural forces, including water-action and trampling will be described at the end of this section.

The location of the edge damage and the nature of the stress that caused them is very important because breaks or micro-flakes on edges of tools reflect use angles, force directions, relative force, and other details of function. Breaks on retouched tools that indicate failure during bending, twisting, impact, or thermal stress are strong evidence for the functions of tools (Collins 1993:90). Therefore, the pattern of use-breakage of stone tools provides information about how tools are damaged through use. Thus, the shape and distribution of these breaks are generally used to distinguish functional variations. The mechanics of use-wear breakage is important to understanding breakage pattern on edge. The principles applied to the breakage of fine-grained brittle solids on a large scale such as flint knapping can be applied to damage to tool margins caused by use on a miniature scale. As a result, flakes caused by use are referred to as microflakes. Major characteristics of microflaking is that the edge of the tool which contacts the worked material plays a role as a striking platform, while the worked material as indenter. The basic principle of fracture mechanics applied to flint knapping will be described in the next chapter. In general, microwear analysis emphasizes both microwear polish (Anderson 1980; Fedje 1979; Lawn and Marshall 1979) and microwear breakage. Analysis based on microwear polish or abrasion is problematic because it is very hard to discriminate between different types of microwear polish (Yerkes and Kardulias 1993:103). As a result, microwear analysis based upon microwear polish has been rejected (Newcomer *et al.* 1986, 1988). The present study focuses instead on microwear breakage or scar patterns.

The Nature of the Worked Material. In use-wear analysis, important variables that affect the formation of edge-damage are the mode of action and the nature of the worked material. The nature of the worked material can be recognized on the basis of the morphological characteristics of the microflake scars as variation in hardness, friction, and resistance of the substance is correlated with variation in size, shape and sharpness of edge of the microflake scars (Tringham *et al.* 1974:188). When a tool contacts with a substance and force is applied, the worked substance often damages the tool margin. Lithic analysts have referred to the worked substance as the "indenter." Depending on its properties, the worked substance can act as a blunt or a sharp indenter, which leads to different fracture. The development of the edge-damage on a tool is largely depending out upon the size of the contact area on the substance. Soft worked material such as animal hide and meat act as blunt indenters and produce only scalar shaped scars. Scalar shaped scars are scars that tend to become wider as they proceed in from the edge. In this case, the mean indention pressure, applied load divided by the contact area, is not enough to fracture the material because the contact area is relatively large. As a result, the initiation of the bending fracture is very high and that of cone-like fracture is very low. Scars produced by working soft material appear to be slight nibblings. The contact with this wide area often causes a greater occurrence of smooth abrasive wear with polish appearance and striations, although this abrasion is hardly visible without special coating at magnification below 100 x. This abrasion usually occurs over a large area of the surface of the tool. If the hardness of worked material is medium between soft and hard worked material, the occurrence of bending fracture is higher than that of hard substance because of a relatively wide contact area with the tool although it provides considerable resistance. In general, woody substance such as softwood like pine or hard wood like oak is considered medium. However, the occurrence of point-cone fracture is fairly frequent on a small scale relative to that of hard material. The working edge angles of tools are generally more acute than for hard materials. Various species of wood material also produce a variety of forms of scalar scars. If the material is hard wood, it yields step scars. Abrasion on the tool is confined to the edges of the scars but intrude farther into the scar than is the case with hard material. Like soft worked materials, the abrasion on the edge is not visible without coating or very high magnifications. If the worked material is harder than woody substances, such as antler and bone, edge wear is characterized by cone-like fractures and a crushing such as hinge and step fracture although initial removals are characterized by large and feather termination because this type of worked materials allows only a small contact area and concentrate the force on that area. Abrasion occurs more frequently and more rapidly on the tool edge relative to other materials. If bone is a worked material, the form of a dull polish appears, while antler tends to produce more striations. (Lawn and Marshall 1979:64-65; Lawrence 1979:115-119; Tringham *et al.* 1974:189-191). As the worked materials are harder, they act as a sharp indenter and the size of the contact area between the implement and the substance decreases. As the size of the contact area decreases, the force exerted on the area increases. As a result, the incidence of edge fracture also increases.

The Modes of Action. The pressure applied to a tool appears at an infinite number of angles to the edge yet it leads to the formation of only two types of fracture. If the force is oblique to the edge, it will bend the tool material beyond its elastic limits and may result in a bending fracture. If the force is directly toward the main body of the tool and approximately bisects the edge, it may result in a point-cone initiation on a Hertzian cone principle. In this case, the initiation of fracture appears either close to the indenting mechanism or at varying distances from it (Odell 1981:198-199). During use, the pressure applied to tool varies according to the type of work being undertaken. Ethnoarchaeological studies reveal that when the action of

scraping is involved, the most common morphological attributes found are a convex bit in plan view and a plano-convex lateral cross-section. If the transverse action across a tool edge is applied to planar surface, that is, the planar surface is oriented downward; the resulting scars are contiguous to one another, feather-terminated and often large in size as several chips are removed from the convex surface. When the convex surface is oriented downward, scars are spaced noncontiguously, step or hinge terminated and small in size with the removal of fewer chips from the planar surface (Odell 1981:200-201). As edge wear implies that the orientation of use-flakes is the same as the direction of force applied to the edge, tools served as a scraper show the flake scars perpendicular to the edge (Lawrence 1979:118). When a motion of shaving is involved, scars are usually removed from the trailing surface with bending fractures. The most frequent scar pattern is of the rolled over, feather-terminated variety. In sum, both scraping and shaving are transverse action to the edge of the tool. As these modes of action related to only one surface of the flake receiving pressure from the worked material, the surface opposite the one that has direct contact with the worked material carries microflakes. Compared to the longitudinal action, the scars are densely distributed in a continuous line along a smaller part of the edge, and have a much smaller range of shapes. The scars are regular in size. The major scar pattern by transverse action is semicircular when scalar, or step scars (Tringham *et al.* 1974:188-189). Unlike scraping and shaving, the action of whittling often produces contiguous row of feather-terminated removals on the lower surface and a much less abundant and contiguous array of hinge fractures on the upper surface (Odell 1981:202).

Cutting/sawing movement usually produces the flake scars on the edge oriented either obliquely or longitudinally to the edge rather than perpendicular to it. (Lawrence 1979:118). The flake scars created by the cutting/sawing action are characterized by a marked denticulation of the tool margin (Odell 1981:203). As both cutting and sawing is longitudinal action, the pressure is directed more or less equally along the intersections of the two planes of the edge. An uneven distribution of the scars along the worked edge differentiates cutting action from sawing action which produces scars in greater density overall along the edge to an equal degree on both surfaces. As the work proceeds increased edge angle and friction result in the edge abrasion on the both surfaces of a tool. When the tool is held at 90 $^\circ$ to the worked material, striations run parallel to the edge. With the decrease in angle, striations become diagonal (Tringham *et al.* 1974:188). Slicing/carving produces very similar pattern to those created by the cutting and sawing activities. However, slicing/carving action produces more damage to one surface than the other (Odell 1981:203-204).

Tools used for the action of boring or drilling have bits composed of a series of edges that intersect with the worked material at one or more points. The combination of vertical and rotary pressures to the drill bit produces torsion, which may lead to the breakage of tool edges or of the entire bit. Stresses including torsion can be easily distinguished from the breakage of an entire drill bit by observing the orientations of the break and the hangnail, projection resulted from the stresses. Microfractures located on the sides and the tip of a borer are difficult to recognize, however, because fractures on these parts have been caused either totally or in part by torsion. Damage to the tip of a borer is distinguished from damage to the sides. Damage to the tip usually consists of crushing and abrading because of the pressure oriented toward the tip. Abrasion appears on the edges and over much of the surface of the tool in the form of polish and striations. The striations are oriented perpendicularly to the point (Odell 1981:205; Tringham *et al.* 1974:189).

Tools involved in chopping often show not only damage from scarring, but also abrasive wear in the form of polish and striations. Fracturing becomes more extensive if the tool is twisted in the object material after contact. As edge angles of tool bit in this category are generally large, bending forces are initiated close to the point of contact causing the removal of hinge- and step-terminated scars. These scars occur on both surfaces of an axe bit because bevel is relatively symmetrical and forces directed to one surface is equal to the other. However, parts of the edge protruded in plan-view may show more fractures than other parts of the margin. This is because the projected portions are the first part to make contact on each stroke and then, receive the brunt of the force on a relatively restricted segment of the edge, or because their orientation to the applied force creates friction and tiny striking platforms that accelerate the removal of scars from the opposite surface (Odell 1981:206) (Table 34).

Edge Wear Caused by Accident or Nonhuman Agency. As mentioned above, edge wear is not only caused by human use but also by other agencies. These agencies can be divided into several categories. Natural agencies include water-action and trampling by people or hoofed animals. Post-depositional factors such as tramping by the prehistoric occupants or archaeological recovery and storage techniques can also produce damage.

Edge Wear Caused by Intentional Retouch. The intentional retouch on stone artifacts also causes damage. In this case, it is not easy to distinguish from use-wear. However, intentional retouches have several criterion. Compared to use-wear, this retouch tends to be larger, more invasive, and more regularly placed. On the other hand, use-wear is usually smaller and less regularly spaced, although it is often concentrated on projecting parts of the edge. If the use-wear appears on a retouched edge, it tends to nick, crush. Despite these various damage agencies, scars produced by non use-wear are very similar to each other. These scars are, in general, distributed at random along the entire perimeter of the flake without localization of scarring. The orientation of the scars is random, with no standardization of scar size or shape except for the those by trampling. Scars caused by trampling appear only on one surface, which is exposed to the tramplers and shows a marked elongation (Odell and Odell-Vereecken 1980:96-97).

Stylistic Analysis

Although classification of the lithic archaeological record has become a fairly refined, the actual cultural significance of these classification is still obscure.

According to Sackett (1982:60), approaches to style in lithic archaeology can be grouped into three types. They are standard, isochrestic, and iconological approaches. All these three approaches are based upon the concept: style is related to a highly specific and characteristic manner of doing something that is always peculiar to a specific time and place. Because of the culture-historical context of its manufacture and employment, style is conventionally regarded as the most important attribute for distinguishing ethnic identification in the prehistoric record.

The Standard Approach. The standard approach to style in archaeology focuses on time-space systematics, based upon the pattern of artifactual similarities and differences. These patterns are then used to order the archaeological record into culture-historically distinct units of ethnic tradition. With this approach one direct correlation between the degree to which artifacts assemblages relate ethnically is assumed to exist and the degree to which they share similar formal variation expressed by artifact types and type frequencies. Style is ascribed to any aspect of formal variation that can be considered diagnostic of ethnicity in time-space systematics. Style in the standard approach means a label applied to variation rather than a concept used to explain it (Sackett 1977:374-375). As a result, style is employed in a highly ambiguous and inconsistent fashion; there is no coherent body of theory that attempts to define explicitly what is and what it presumably accounts for (Sackett 1982:60-63).

The Isochrestic Approach. The isochrestic approach to style also emphasizes the time-specific systematics but this approach is different from the standard approach. In the isochrestic approach, both function and style are viewed as fully complementary aspects that share equal responsibility for all formal variation observable in artifacts. Thus, they can not be comprehended except in terms of the each other. When style and function have been totally accounted for, all formal variation in the artifacts can be explained. Style in the isochrestic approach equates ethnicity with functionally equivalent choice. Stone artifacts are particularly rich in analogous examples of stylistic-functional duality. This duality of function and style exists at the microlevel of the formal attributes of which they are composed as well as the macrolevel of the assemblages and industries into which they combine as individual artifact types (Sackett 1973:318-321).

G-Young Gang

Table 34. The Mode of Action and Use wear Pattern

Direction of action	Mode of action	Use-wear pattern
Transverse	Scraping	Step or hinge fractures perpendicular to the edge
	Shaving	Feather terminated variety and step fractures perpendicular to the edge
	Whittling	The lower surface: contiguous row of feather terminated removals. The upper surface: less abundant and contiguous array of hinge fractures
Longitudinal	Cutting	Denticulation oriented obliquely or longitudinally to the edge or uneven distribution of scars
	Sawing	Oblique or longitudinal scars in greater density overall along the edge to an equal degree on both surfaces
	Slicing and carving	Similar pattern to those created by cutting and sawing but more damage to one surface
Rotational	Boring	Appearance of crushing and abrasion on the tip; abrasion in the form of polish and perpendicular striation on the edges
	Chopping	Appearance abrasion in the form of polish and striation on the surfaces; hinge and step terminated scars on the contact area with worked material

The Iconological Approach. Style in the iconological approach is equated with specific elements of non-utilitarian formal variation which function symbolically as a kind of "ethnic iconography", which symbolically expresses social information. Thus, iconological approach states specifically wherein style resides: it is to be found in either non-utilitarian objects or adjunct form on utilitarian objects. In contrast to the isochrestic approach, the iconological approach offers an inside and essentially functional view of style itself. The iconological approach to style is a highly specific model of how non-utilitarian objects and adjunct form on utilitarian objects function in the societal and ideational realms. The approach emphasizes the facts that as stylistic elements constitute an iconography of social relationships; certain kinds of form are chosen and used stylistic because they convey social information. If form does not have as its primary function the symbolic expression of social information, it is not stylistic in nature but functional (Binford 1963). Therefore, the major difference between the isochrestic and the iconological approaches is that in the former approach, style resides potentially in all formal variation, while in the latter it becomes a distinct realm of form divorced from all other realms (Binford 1986:560; Sackett 1982:95, 1986:630).

In the present study, I find stylistic variability on the basis of isochrestic approach. Most of tools in the present lithic assemblage consists of either scrapers or knives which is characteristic of Middle Stone Age in Africa and Middle Paleolithic in Europe. Although there is possibility to find symbolic information that resides in non-utilitarian objects, it is very difficult to find symbolic or idealistic realm on the basis of iconological approach. This is because (1) tools in the present assemblage consist largely of flakes, which are not modified by intention but by use. As a result, stylistic variability is closely related to function; (2) archaeological sites which produced the present lithic assemblage are purely lithic sites. In other words, both stylistic and functional variability are also affected by techno-economic constraints such as raw material constraints, and different degrees of artifact-reduction intensity (Rolland and Dibble 1990:480). Therefore, the stylistic variability in the tool assemblage should be analyzed in terms of form and function.

Research Design

There are four primary factors that influence or limit the shape that a stone artifact may assume (Whittaker 1994:270). These factors include: (1) the raw material of

96

which the tool is formed, (2) the technology used in its manufacture, (3) the function of the object, that is, the purpose or purposes to which it is to be put, and finally, (4) the style of characteristic mode of tool form current among the people making the object. All four of these factors were systematically considered in comparing the lithic assemblages from the various components at the Shurmai (GnJm 1) and Kakwa Lelash (GnJm 2) rockshelters. To achieve this goal, the lithic assemblages from two sites were categorized as debitage, cores, and tools. For debitage analysis, decortication debitage analysis, tool debitage analysis, and hierarchical debitage analysis were applied to the present study. To reconstruct core reduction sequence, both core assemblage and flake assemblage were analyzed. The former consists of exhausted and unexhausted cores. The latter used in core analysis is composed of the attributes that are also used in the debitage analysis. These attributes include the nature of flake scar and the existence of cortex on exterior flake surface. In tool analysis, both the function of the object and the style of characteristic mode of tool form were focused in the present study. In order to understand the function, edge angles were measured and edge wear was studied. Based upon the information obtained from the analysis of function, style of the tool was analyzed. The importance and application of all attributes or variables for each category in the present lithic analysis are described above in debitage, core, and tool analysis sections. Additional information will be described in this section as well.

Raw Material

Understanding raw material characteristics is important because fracture mechanics and tool type are dependent on raw material. For instance, it is very difficult to produce a thin biface on basalt or quartzite. Stone quality suitable for knapping will be described in the following chapter. Analysis of raw material also provide us with a key to understanding the range of our ancestors' environmental exploitation and raw material procurement through those resources being exploited through use of the lithic technology and the variability of the raw materials used in lithic manufacture in terms of accessibility, physical properties, and suitability for the processes of manufacture and the use to which they were subjected. For the debitage analysis, the category of raw material in the present study includes eight separate stone types and one miscellaneous category (see Appendices A-C). The eight types include the four major material types of such as basalt, chert, obsidian, and quartz and the four minor material types of rhyolite, quartzite, sandstone, and gneiss. The category of chert includes chert, jasper, and chalcedony. Although further distinctions are possible between chert, jasper, and chalcedony, I decided to combine them since both chert and jasper are impure varieties of chalcedony, which in turn is a cryptocrystalline variety of quartz and predominantly composed of silica.

The goals of raw material analysis in the present study is (1) to compare raw material of flakes, cores and tools which each unit produced and (2) to seek differences in ratios of cores to flakes and tools to flakes. The type of raw materials used in stone working at the the two shelters will be undertaken with reference to the obsidian source studies for Kenya (Merrick and Brown 1984; Merrick, Brown and Connelly 1990).

Techno-Morphological Analysis

The technology used to produce stone artifacts includes both knowledge and the ability to use techniques and tools (Whittaker 1994:270). Morphology is a method of categorizing stone artifacts based on form alone. However, tools of the same form may be produced by different technologies. A "techno-morphological" study considers both morphology and technique in the analysis of a stone tool assemblage.

The system of classification for the present study is based on the techno-morphological analyses of Bordes (1961), Debénath and Dibble (1993), Fish (1979), and Van Peer (1992). Although all these analyses use attributes of lithic debitage recovered from Middle Paleolithic sites in Europe and Asia, I have used their methods in my analysis for three reasons. First, their methods are widely known, accepted and understood. Second, the techno-morphological analysis of African Middle Stone Age lithic assemblages is not as well developed as the foregoing. Third, both the archaeological record and the practical experience of lithic technologists indicate that certain technical attributes are associated regardless of the site location and period. My analysis modifies three methods mentioned above and uses attributes relating to broad range of flake techno-morphology. The attributes used in the present study are discussed in the following section.

Debitage Analysis. As noted above, debitage is a broad category that includes all flaked stone materials not recognized either as cores or as tools. Debitage is very important because critical information useful in reconstructing prehistoric lithic technology and patterns of human behavior can be obtained from it. The present study modifies Bordes (1961), Debénath and Dibble (1993) and Fish (1979). Attributes are described in the order of their occurrence on the coding sheets. A sample of my coding sheet is attached in Appendix A.

Completeness. All debitage was classed either as complete, broken, fragment, or as debris. Of these, both complete and broken flakes that contain striking platform and bulb were analyzed because the platform type is very important attribute of the Middle Stone Age stone artifacts.

Condition. Most of stone artifacts in the present study exhibit one of three conditions. They are either burned, patinated, or unaltered. Heat-treatment of stone improves stone quality for knapping. Heat-treatment works for some kinds of stone such as chert by rendering them less grainy, smoother in texture, more brittle and easier to flake (Whittaker 1994:72). A more detail explanation on of heat-treating is provided in the Chapter six. Patina is a generic term used to describe the chemically altered surface of stone. Patinas are produced by many deterioration processes such as weathering. Patination is caused by chemical, physical, and biological agents. Of these, chemical weathering by water is the most important. In archaeological sites different deposition history cause distinct patinations. Cultural formation processes such as reuse and recycling also affects patination of stone artifacts (Schiffer 1987:152-154, 274).

Maximum Length. This is a measure of the longest dimension of a flake. In order to systematically measure

the maximum length of a flake, it is necessary to draw the smallest rectangle that completely enclose the entire flake. The axes drawn from the striking platform and the distal end of the flake are used for orientation on one axis of the smallest rectangle and axes drawn from right and left flake edges are used on the other. The flake is laid flat on a sheet of paper and two lines intersecting the distal and proximal ends are drawn parallel to one another. Then two lines that intersect the flake edges at their widest points are drawn parallel to one another. The measurement is taken from the long axis of the rectangle and made in millimeters using calipers. Figure 26 A-B shows the maximum length of the flake.

Maximum Width. This is a measure of the widest

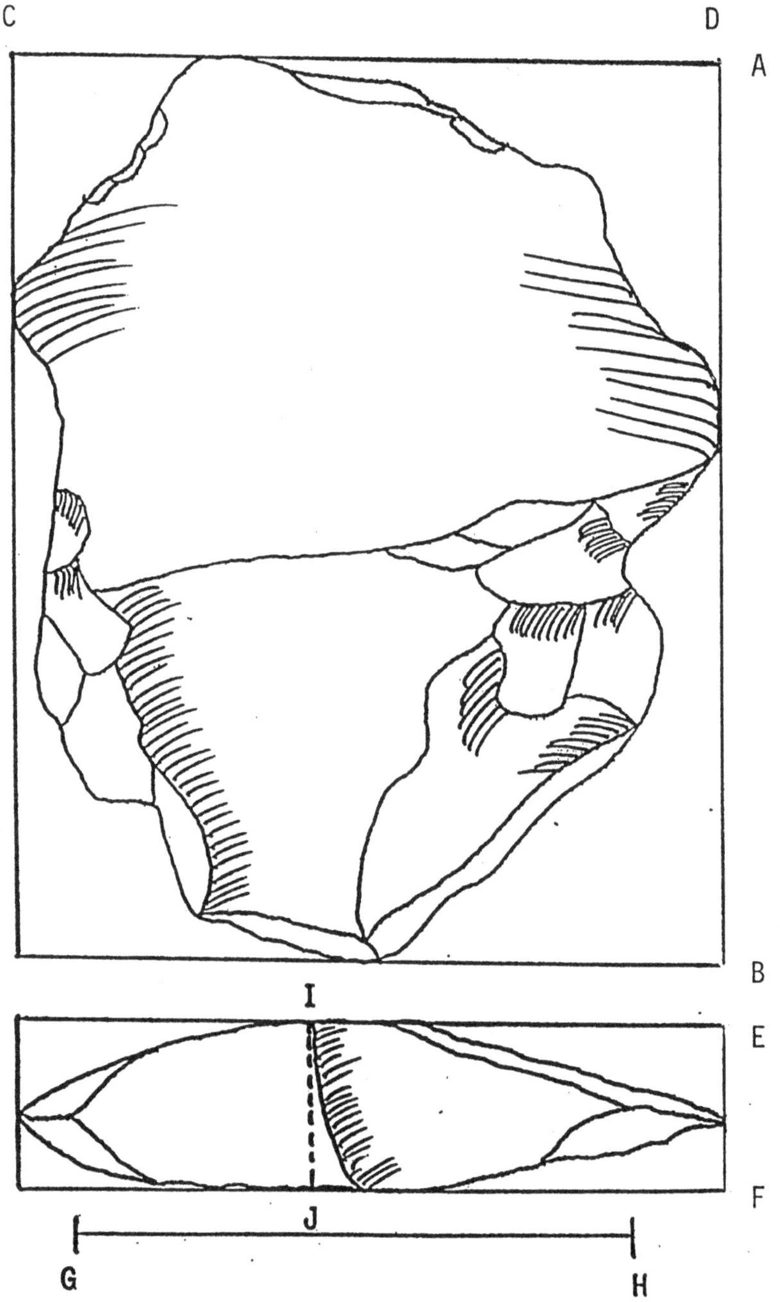

Figure 26. Measurement of Flake Attributes in Debitage Analysis.

dimension of a flake. The method of flake orientation is same as described for the maximum length. Right and left lateral edges of a flake are used for the axes. The measurement is made in millimeters using calipers. Figure 26 C-D shows the maximum width of the flake.

Maximum Thickness. This is a measure of the thickest dimension of a flake. The method of flake orientation is the same as described for the maximum length. The maximum depth between the interior and exterior surfaces is measured. In general, the depth between the center of the bulb of percussion on interior surface and exterior surface is the thickest dimension in the flake.

Striking Platform Width. This measure is made at the widest point on the striking platform. It is made in millimeters using calipers. Figure 26 G-H shows the striking platform width.

Striking Platform Depth. This measurement is taken at the center of the point of impact and perpendicular to the axis formed by the striking platform width. It is made in millimeters using calipers. Figure 26 I-J shows striking platform depth.

Angle of the Striking Platform. Flakes have both interior and exterior platform angles. In the present study, the angle of the striking platform refers only to the exterior platform angle. Definitions of each platform angle are described in debitage section. A goniometer is used to measure exterior platform angle. A goniometer is a device for measuring solid angle that is widely used in crystallography. The armature of a goniometer is placed at 90° position. The striking platform of the flake is positioned perpendicularly placed against the armature with its dorsal surface facing down. The armature is tilted to measure the angle formed by the intersection of the platform surface and the dorsal surface of the flake. Measurement of exterior platform angle is illustrated in the debitage section (Figure 27).

Striking Platform Characteristics. Seven platform types and one miscellaneous category are recognized in this study. All eight types are illustrated in Figure 28. These include:
(1) Cortical - the striking platform retains cortex. Figure 28.a
(2) Plain - the striking platform is characterized by a single facet and shows a smooth and previously flaked surface. Figure 28.b
(3) Dihedral - the striking platform exhibits two facets that intersect at a relatively sharp angle. Figure 28.c
(4) Polyhedral - the striking platform exhibits three or more facets along the surface. Figure 28.d
(5) Faceted - the striking platform is characterized by the intersection of three or more facets along the surface. Figure 28.e

(6) Reworked - the striking platform was retouched after a flake was removed form the core. Figure 28.f
(7) Broken - the striking platform has been identified as broken off the flake.
(8) Other or Miscellaneous - all flakes of unrecognized striking platform type.

Planform Shape. This is the shape of the exterior surface of a flake when it is oriented vertically along the axis of the piece with the proximal end down. Ten types of planform shape are recognized including: (1) divergent, (2) parallel, (3) convergent, (4) medially expanded, (5) irregular, (6) broken, (7) skewed, (8) miscellaneous, (9) oval, and (10) round. Here, parallel shape is assigned only to blades. Figure 29 shows illustration of the nine planform shapes except for miscellaneous.

Flake Scar Count. The number of flake scars evident on exterior surface. Only scars larger than 2 mm are counted.

Flake Scar Orientation. All directions are identified on the basis of the striking platform of a flake when the flake is oriented vertically along the axis of the piece with the proximal end downward and the distal end is upward. Eleven of flake scar orientation types are recognized. These include: (1) parallel unidirection, (2) perpendicular unidirection, (3) irregular unidirection, (4) divergent unidirection, (5) convergent unidirection, (6) parallel bidirection, (7) irregular bidirection, (8) centripetal or radial direction, (9) random, (10) no pattern, (11) undetermined. Figure 30 shows flake scar orientation.

Flake Terminations. Flake termination refers to the distal end and edges of a flake. Five types of flake terminations are recognized. Technical explanation of various types of formation of flake terminations is provided in the following chapter. These are:
(1) Feather - a flake with edges and a distal end that is very sharp.
(2) Hinge - the distal end of a flake which turns sharply upward, forming a rounded hinge on the end of the flake.
(3) Step - the distal end of a flake characterized by a right-angle break. This generally indicates the flake is broken.
(4) Overshoot - the distal end of a flake contains part of the end of the core that was located beyond the opposite margin.
(5) Broken - flakes that do not have distal part because of breakage. In contrast to the step termination, broken distal does not show right-angle break.

Presence or Absence of Cortex. Presence of cortex on the exterior flake surface is accounted: (1) present and (2) absent.

Placement of Cortex. The placement of the cortex is determined by when the flake is oriented vertically along the axis of the piece with the proximal end down. Seven

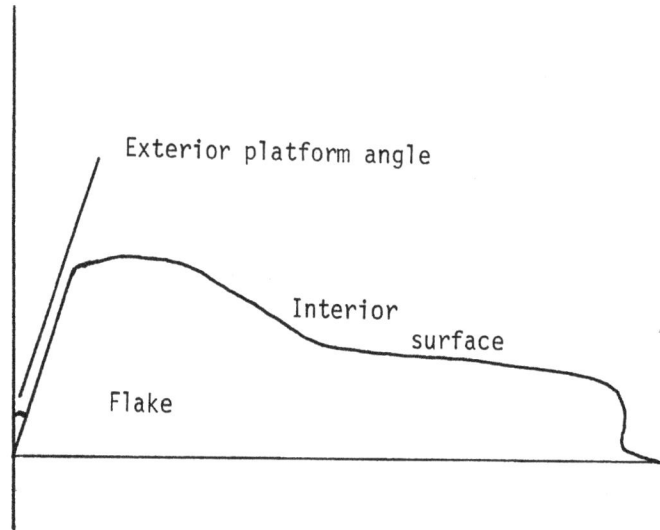

Figure 27. Measurement of the Angle of the Striking Platform.

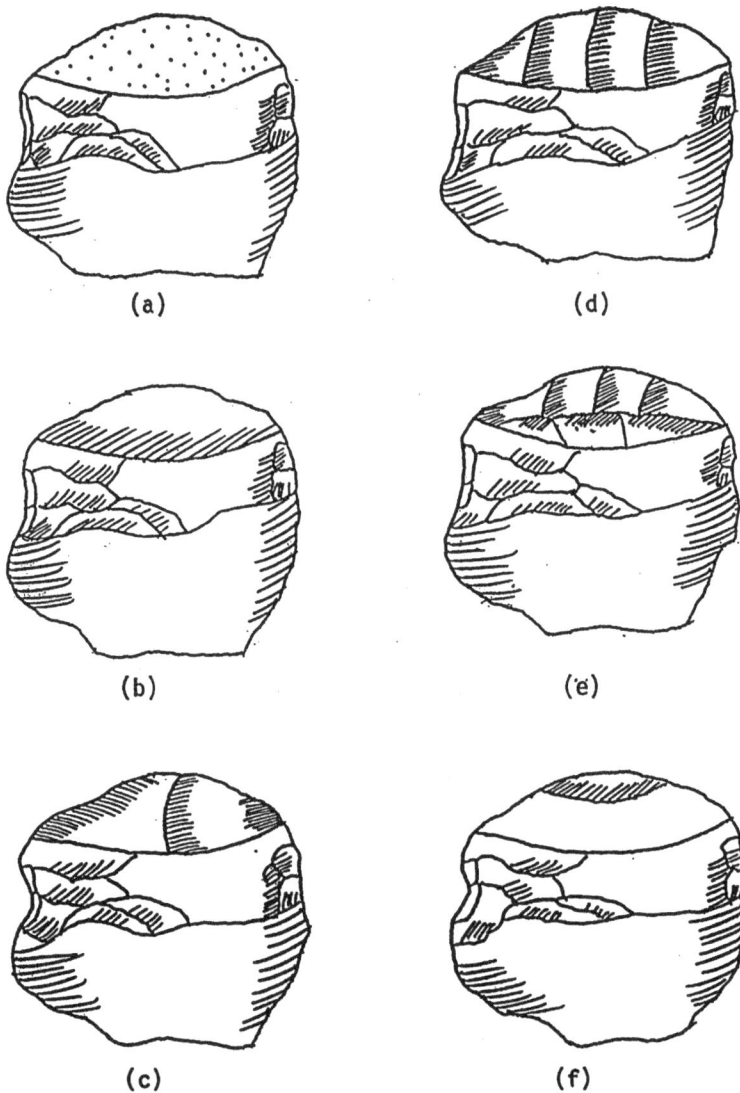

(a)

(d)

(b)

(e)

(c)

(f)

Figure 28. Types of Platform.

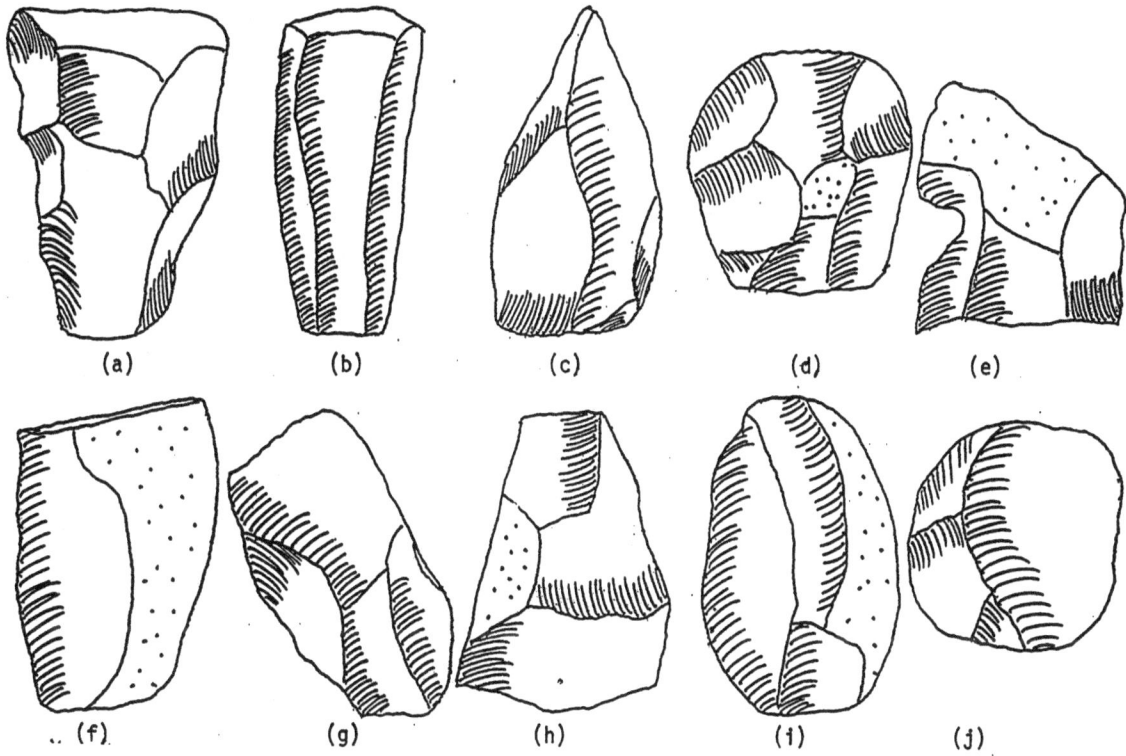

Figure 29. Planform Shapes.

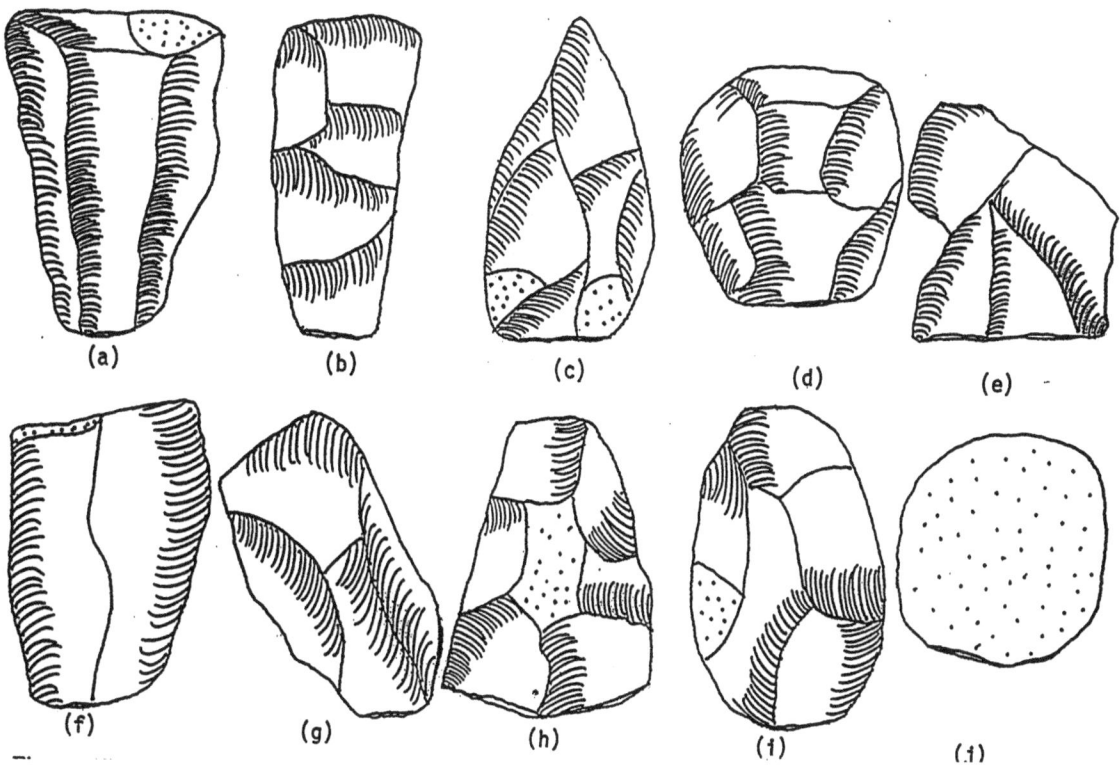

Figure 30. Flake Scar Orientation.

types are recognized. Identified seven categories include: (1) proximal, (2) distal, (3) left edge, (4) right edge, (5) dorsal, (6) center, and (7) all around the edge.

Estimate of Percentage of Cortex. This is a subjective estimate of the amount of cortex present on the exterior surface of the flake. Three categories are recognized including: (1) primary - flakes which have dorsal surface covered with cortex (2) secondary - flakes which contain some cortex on their dorsal surface, and (3) tertiary - flakes without cortex.

Core Analysis. In the core samples from the assemblages under consideration, many of them have two convex faces opposite to each other. The present study modifies Baumler (1987) and Van Peer (1992). Attributes are described in the order of their occurrence on the coding sheets.

Dorsal Surface. Cores are first divided into dorsal and ventral surfaces. If there is cortex on the surface, it is called dorsal surface. However, some cases the distinction between dorsal and ventral is not clear. In this case, the distinction is subjective. Dorsal surface of the cores are divided into four sectors, each equal to one circle quadrant. The center of the core is placed in the circle midpoint. The central core axis is placed along a vertical line delimiting two circle segments. The quadrants are placed at 90 degrees to the vertical axis. For each sector, the number of scars and the number of remaining negative bulbs are counted. In addition, total number of flake scars, that is, flake scars on all four quadrants, and the total number of negative bulbs are recorded.

Ventral Surface. The method of analysis is as the same as the one described for the dorsal surface. Figure 31 shows the deposition of cores in quadrants.

Maximum Length. This is taken along the axis of symmetry of the core. The measurement is made in millimeters using calipers. Figure 32 A-B shows the measurement of maximum length.

Maximum Width. It is made in one vertical plane, perpendicular to the maximum length. The measurement is made in millimeters using calipers. Figure 32 C-D illustrates maximum width.

Maximum Thickness. It is taken along the maximum distance between the dorsal and ventral surfaces. The measurement is made in millimeters using calipers. Figure 32 E-F shows maximum thickness.

Flake Scar Pattern. Based upon flake scar directions, four types are recognized. These are (1) unidirectional, (2) bidirectional, (3) radial, and (4) random.

Completeness. Cores are grouped into exhausted and unexhausted. In either case, if the core is complete it is

categorized as (1) whole and if the core is identified as broken, it (2) fragment.

Material Type. Most of cores recovered from the sites are made of four types of raw materials. They are (1) basalt, (2) chert, (3) obsidian, and (4) quartz.

Condition. Four types of conditions are identified. These are (1) burned, (2) battered (3) patinated, and (4) miscellaneous. Unaltered raw material is included in the last category.

Tool Analysis. In this final section research design for the techno-morphological aspect of tools in the present assemblage. Attributes are described in order of their occurrence on the coding sheets.

Tool Type. Tools were grouped into three types including: (1) cobble, (2) flake, and (3) non-flaked or ground. Of these groups, non-flaked or ground tool type includes such as hammer stone and mano.

Material. All tool materials recovered from the sites are made of five types of raw materials. They are (1) basalt, (2) chert, (3) obsidian, (4) quartz, and (5) quartzite. However, one miscellaneous group was made for materials that are not dominant in the present assemblage.

Condition. Seven conditions were suggested. These are (1) burned, (2) patinated, (3) battered, (4) water worn, (5) no observable pattern, (6) soil polished, and (7) miscellaneous. Of these, first two was explained in the debitage section. Battered condition is created for the tools that were used as chopper or axes. Water worn condition for the tools weathered by water, while any observable pattern refers to no alteration. Although soil polish is included in patination, I try to distinguish polish from other patinated conditions.

Planform Shape. Planform shape is described in the debitage section. Here, only the order and categories are different from those in the debitage section. Seven types of planform shape are recognized including (1) divergent, (2) parallel, (3) convergent, (4) medially expanded, (5) irregular, (6) oval, and (7) round. Of these, irregular group includes unidentified planform shape also.

Completeness. Each tools made of a flake is analyzed on the basis of the presence of proximal portion of the flake. They are (1) whole, which contains both proximal and distal parts of a flake, (2) distal fragment, (3) medial fragment, (4) proximal fragment, and (5) undetermined. If the tool is made of either cobble or non-flaked material, both whole and undetermined groups were applied to.

Type of Modification. Eight types were recognized. These include: (1) Percussion, (2) pressure, (3) percussion and pressure, (4) use wear, (5) pressure and use wear, (6)

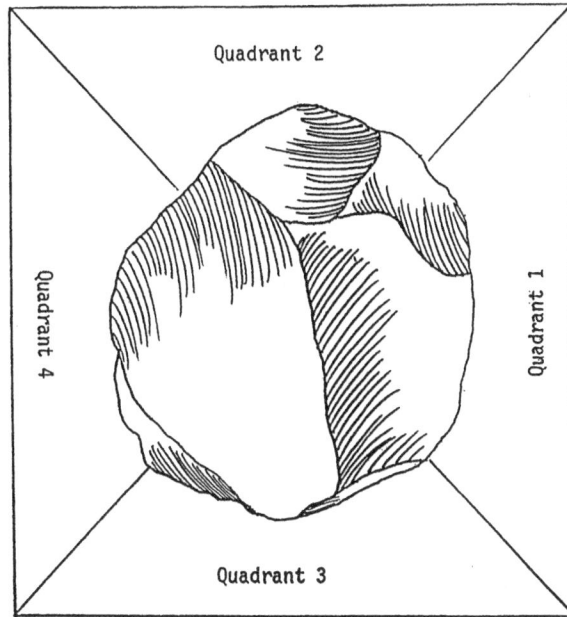

Figure 31. Deposition of a Core in Quadrants.

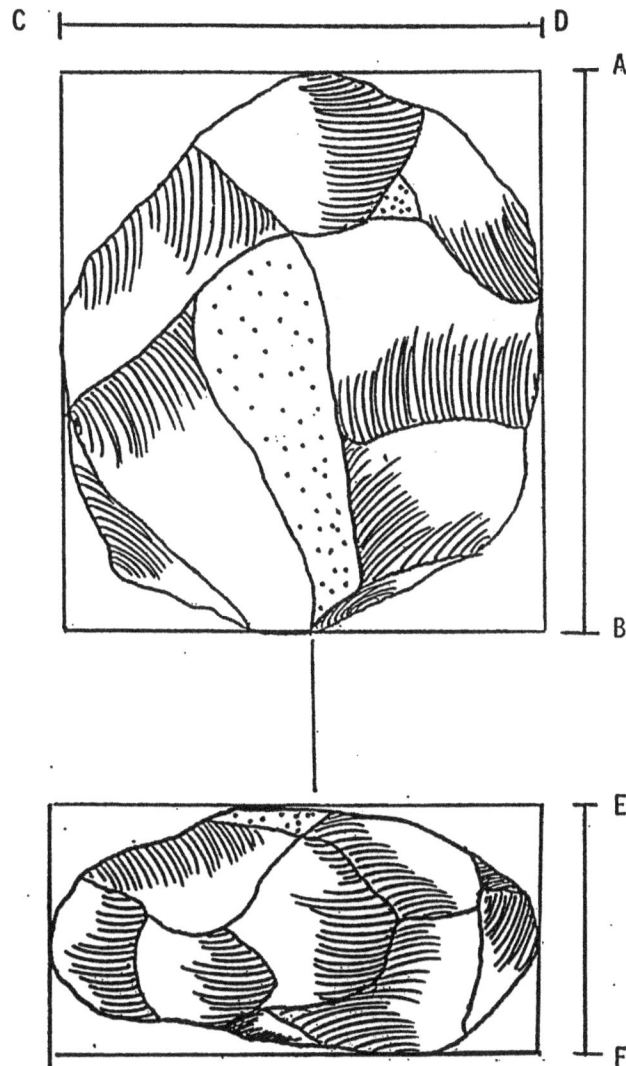

Figure 32. Measurement of Core Dimension.

percussion and use wear, (7) percussion, pressure, and use wear, and (8) undetermined.

Surface of Modification. Tools made of only flake is accounted: (1) exterior surface, (2) interior surface, and (3) both exterior and interior surfaces.

Location of Modification. This is the placement of retouch. They are: (1) proximal, (2) distal, (3) medial, (4) around the entire edge, (5) miscellaneous.

Retouch Type. Eight types of retouch, which most widely appeared in Middle Paleolithic lithic assemblage, are suggested for this analysis. They are:
   (1) Scalar - retouch scars tend to become wider as they proceed in from the edge.
   (2) Step - retouch scars tend to end in either step or hinge fracture.
   (3) Quina - further development of step retouch on tool.
   (4) Parallel - the margin of the retouch scars are parallel. This retouch type is generally obtained by pressure technique.
   (5) Nibbling - Extremely light but continuous retouch scars, which generally appear at the extreme edge of the tool.
   (6) Backing - abrupt and continuous retouch.
   (7) Notching - the production of a single, relatively deep concavity on a tool margin. This retouch is generally non-continuous.
   (8) Denticulation - a series of contiguous notches or serration along the edge of a tool. Figure 33 shows the types of retouch.

Dimensions for Maximum Length, Maximum Width, and Maximum Thickness. All the methods applied to measurement of these dimensions of tools are also described in debitage section.

Butt Type, Butt Width, Butt Thickness, and Platform Angle. All the methods applied to measurement of these dimensions of tools are also described in debitage section.

Functional Analysis

For the functional analysis, edge angle and use-wear were studied.

Edge Angle. The measurement of edge angle varies. The method of edge angle in the present study modified that of Wilmsen (1967). The measurement of edge angle of the tool was taken from the angle that is formed between the extreme edge and spine or ridge on the tool.

When the retouched part on the tool is more then 1 cm, the measurement was made every 1 cm because of the irregular shape produced by use or retouch. The measurement was made by a goniometer. Figure 34 shows the measurement of edge angle.

Use-Wear. Use wear can be examined by either the low-power or the high-power techniques. Although both techniques are useful for the use wear analysis, advocates of each approach have suggested benefits of their method and pointed out the problems with the other technique. The low-power method pioneered by Semenov, in general, uses one microscope with capabilities up to 10 x to 100 x. The major problems of the low-power technique questioned by the adherent of the high-power technique includes that the damage caused by intentional retouch and manufacture damage to edges produced during the original removal of the piece and natural movement of soil sediment are extremely difficult to distinguish from use wear. However, the low-power method has merit. First, preparation of the object with chemicals or metallization is usually not necessary. If the low-power approach requires preparation, it is very minimal specimen preparations, so analysis can proceed rapidly and the analyst can examine larger samples. Second, stereo microscopes are relatively inexpensive compared to the metallurgical incident light microscopes needed for the high-power approach (Keeley and Newcomer 1977; Odell and Odell-Vereecken 1980; Tringham *et al*. 1974).

The high-power technique developed by Keeley requires an incident light source in a metallurgical microscope. Examination is done at magnifications between 50 x and 500 x. The major advantage of this method is that a series of micropolishes can be visible that are specific to the material that was worked. The conjunction of micropolish patterns, striations, and edge damage allows the analyst to identity how the tool was used. However, the high-power approach requires careful cleaning of artifacts prior to study. In other words, because of this initial processing time and the need to scan the artifacts at several different magnifications, this approach needs long period of time. In addition, high-power microwear analysis makes sample selection a critical element. Only fine-grained cryptocrystalline lithic materials such as chert and some igneous rocks can be examined with the high-power technique (Keeley 1980; Yerkes and Kardulias 1993:101-103).

Although the results from the high-power studies have greater reliability due to the finer resolution of wear traces that the method allows, use wear analysis in the present study will be examined by low-power approach. This is because first, most stone artifacts in the present lithic assemblage consist of basalt which are seriously weathered. It is doubtful whether the cleaning process can help these seriously weathered material. Second, the low-power technique is useful for the use-wear on basalt. Although Keeley and Newcomer (1977) suggest that it is not easy to recognize use wear caused by non-human use from that by human, experiments by Tringham *et al* (1974) demonstrate the differences between these two wear pattern and they could examine this pattern by the low-power technique. Therefore, the present study employs the low-power technique rather than the high-power technique.

(a)   (b)   (c)   (d)

(e)   (f)   (g)   (h)

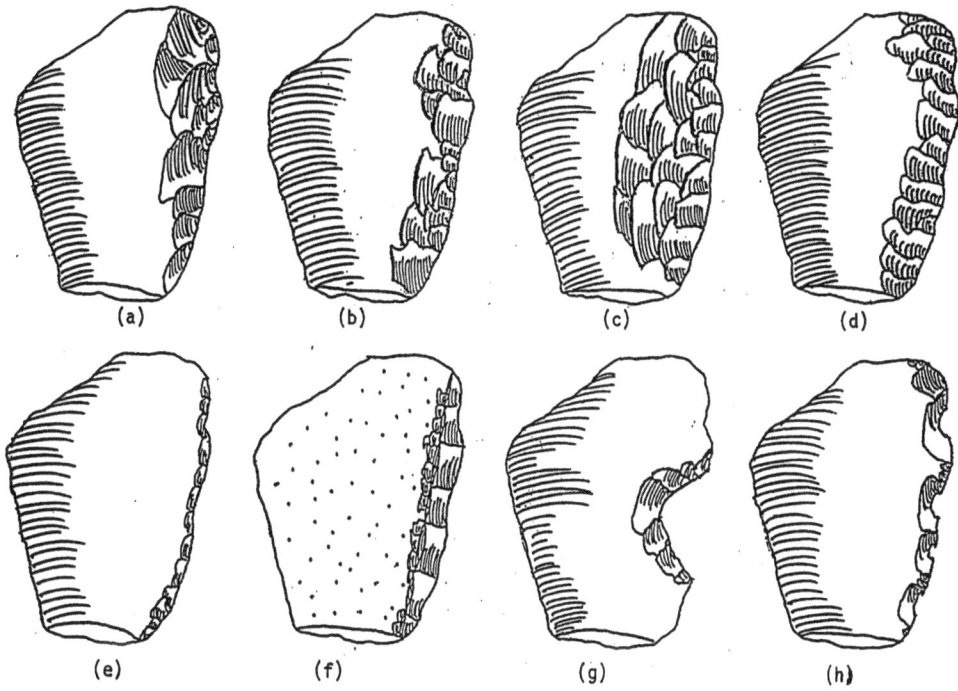

Figure 33. Types of Retouch.

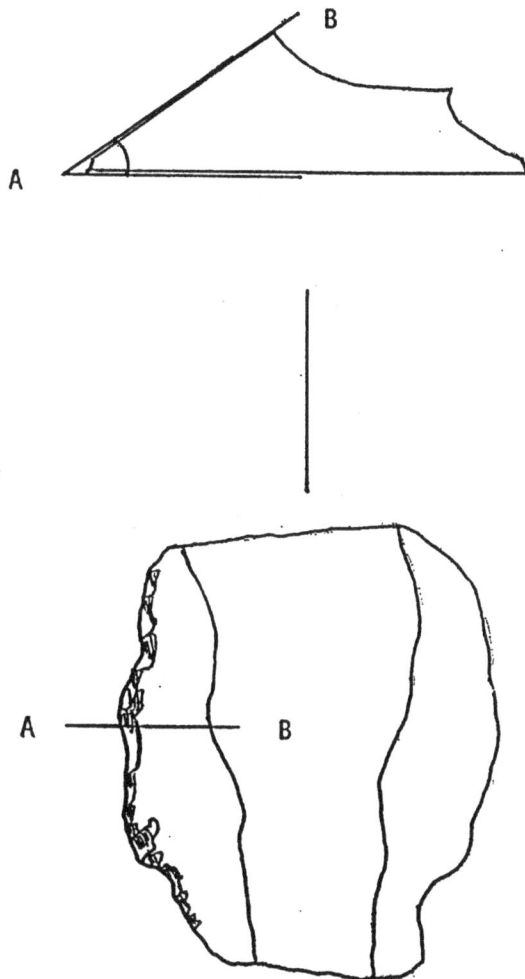

Figure 34. Measurement of Edge Angle.

Equipment for the Use Wear Analysis. In the present study most of tool material analyzed for the functional variability is the material which is kept in Texas A&M University. Many other tool material analyzed and kept in the National Museum of Kenya in Kenya cannot be analyzed. The equipment for the use wear analysis in the present uses a Nikon Stereo microscope with a magnification range of 0 - 60 x at the Palynology Lab of Anthropology Department. An Erinreich Mk II Fiber Optic Light source was used in conjunction with microscopes.

In the present study, the stylistic analysis is based upon the isochrestic approach. This approach is already described in the tool analysis section.

Appendix C
Lithic Analysis Forms

UNMODIFIED LITHIC FLAKE ANALYSIS FORM

Site #:   Date: Name:

Sack #:   Grid Square: Stratum/Level: Feature #:   Specimen No.:

Material type: 1. Basalt   2. Chert   3. Obsidian   4. Quartz
                 5. Rhyolite   6. Sandstone   7. Tuff   8. Gneiss

Completeness: 1. Whole   2. Broken with bulb   Condition: 1. Burned   2. Patinated   3. Unaltered

Size (>2 cm.):   Maximum Length: Maximum Width: Maximum Thickness:
                 Butt Width (Platform)   Butt Thickness (Platform):   Platform Angle:

Butt Type: 1. Cortical   2. Plain   3. Dihedral   4. Polyhedral
                 5. Faceted   6. Re-worked   7. Broken   8. Miscellaneous

Planform Shape: 1. Divergent   2. Parallel   3. Convergent   4. Medially Expanded   5. Irregular
                   6. Broken   7. Skewed   8. Miscellaneous   9. Oval   10. Round

Flake Scar Count (> 2 cm):

Flake Scar Orientation (Dorsal):   1. Parallel Unidirection   2. Perpendicular Unidirection
3. Irregular Unidirection   4. Divergent Unidirection   5. Convergent Unidirection
6. Parallel Bidirection   7. Perpendicular Bidirection   8. Irregular Bidirection
9. Centripetal/Radial   10. Random   11. No Pattern   12. Undetermined

Flake Terminations: 1. Feathered   2. Hinge   3. Step   4. Plunging   5. Other   6. Broken Distal

Cortex: 1. Present   2. Absent

Placement of Cortex: 1. Proximal   2. Distal   3. Left Lateral
4. Right Lateral   5. Dorsal   6. Center   7. Around the Edge

---

107

LITHIC CORE ANALYSIS FORM

Site #:   Date: Name:
Sack #:   Grid Square: Stratum/Level: Feature #:   Specimen No.:

| Dorsal Surface: Total Number of Flake Scars | Quad 1: | Quad 2: | Quad 3: | Quad 4: |
Total Number of Flake Scars (All Quadrants):

| Total Number of Negative Bulbs Each Quadrant: | Quad 1: | Quad 2: | Quad 3: | Quad 4: |

| Ventral Surface: Total Number of Flake Scars: | Quad 1: | Quad 2: | Quad 3: | Quad 4: |
Total Number of Flake Scars (All Quadrants):

| Total Number of Negative Bulbs Each Quadrant: | Quad 1: | Quad 2: | Quad 3: | Quad 4: |

Maximum Length: Maximum Width: Maximum Thickness: Weight:

Index of Elongation (Length) (Width):   Index of Flattening: (Width) (Thickness):

| Flake Scar Pattern: | 1. Unidirectional | 2. Bidirectional | 3. Radial | 4. Random |

| Completeness: | 1. Whole | 2. Fragment |

| Material: | 1. Basalt | 2. Chert | 3. Obsidian | 4. Quartz |

| Condition: | 1. Burned | 2. Battered | 3. Patinated | Unaltered |

LITHIC TOOL ANALYSIS FORM

Site #:   Date: Name:

Sack #:   Grid Square: Stratum/Level: Feature #:   Specimen No.:

Tool Type:
Typology:
1. Cobble   2. Flake   3. Non-flaked or Ground (Hammer Stone, Mano, etc.)

Material: 1. Basalt   2. Chert   3. Obsidian   4. Quartz   5. Quartzite   6. Other

Condition: 1. Burned   2. Patinated   3. Battered   4. Water Worn   5. No Observable Pattern
    6. Soil Polished   7. Miscellaneous

Planform Shape: 1. Divergent   2. Parallel   3. Convergent   4. Medially Expanded
    5. Irregular   6. Oval   7. Round

Completeness: 1. Whole   2. Distal Fragment   3. Medial Fragment   4. Proximal Fragment   5. Undetermined

Type of Modification: 1. Percussion   2. Pressure   3. Percussion and Pressure   4. Use-wear Modification
    5. Post-depositional   6. Other   7. Pressure and Use-wear   8. Percussion and Use-wear
    9. Percussion/Pressure and Use-Wear   10. Undetermined

Modification: 1. Exterior   2. Interior   3. Both

Location of Modification: 1. Proximal   2. Distal   3. Medial/Lateral   4. Around the Entire Edge   5. Other

Retouch Type: 1. Scalar   2. Step   3. Quina   4. Parallel   5. None   6. Nibbling
    7. Backing   8. Notching   9. Denticulation

Maximum Length: Maximum Width: Maximum Thickness:

Butt Type (Only Flake Tools):   1. Cortical   2. Plain   3. Dihedral   4. Polyhedral
    5. Faceted   6. Reworked   7. Polyhedral   8. Miscellaneous

Butt Width: Butt Thickness: Platform Angle:

CAMBRIDGE MONOGRAPHS IN AFRICAN ARCHAEOLOGY

No 31   BAR S455, 1988 **Shellfish in Prehistoric Diet Elands Bay, S.W. Cape Coast, South Africa** by W.F. Buchanan. ISBN 0 86054 584 9

No 32   BAR S456, 1988 **Houlouf I** *Archéologie des sociétés protohistoriques du Nord-Cameroun* by Augustin Holl. ISBN 0 86054 586 5

* No 33 BAR S469, 1989 **The Predynastic Lithic Industries of Upper Egypt** by Liane L. Holmes. ISBN 0 86054 601 2 (two volumes)

No 34   BAR S521, 1989 **Fishing Sites of North and East Africa in the Late Pleistocene and Holocene** *Environmental Change and Human Adaptation* by Kathlyn Moore Stewart. ISBN 0 86054 662 4

* No 35 BAR S523, 1989 **Plant Domestication in the Middle Nile Basin** *An Archaeoethnobotanical Case Study* by Anwar Abdel-Magid. ISBN 0 86054 664 0

* No 36 BAR S537, 1989 **Archaeology and Settlement in Upper Nubia in the 1st Millennium A.D.** by David N. Edwards. ISBN 0 86054 682 9

No 37   BAR S541, 1989 **Prehistoric Settlement and Subsistence in the Kaduna Valley, Nigeria** by Kolawole David Aiyedun and Thurstan Shaw. ISBN 0 86054 684 5

No 38   BAR S640, 1996 **The Archaeology of the Meroitic State** *New perspectives on its social and political organisation* by David N. Edwards. ISBN 0 86054 825 2

No 39   BAR S647, 1996 **Islam, Archaeology and History** *Gao Region (Mali) ca. AD 900 - 1250* by Timothy Insoll. ISBN 0 86054 832 5

No 40   BAR S651, 1996 **State Formation in Egypt:** *Chronology and society* by Toby A.H. Wilkinson. ISBN 0 86054 838 4

No 41   BAR S680, 1997 **Recherches archéologiques sur la capitale de l'empire de Ghana** *Etude d'un secteur d'habitat à Koumbi Saleh, Mauritanie. Campagnes II-III-IV-V (1975-1976)-(1980-1981)* by S. Berthier. ISBN 0 86054 868 6

No 42   BAR S689, 1998 **The Lower Palaeolithic of the Maghreb** *Excavations and analyses at Ain Hanech, Algeria* by Mohamed Sahnouni. ISBN0 86954 875 9

No 43   BAR S715, 1998 **The Waterberg Plateau in the Northern Province, Republic of South Africa, in the Later Stone Age** by Maria M. Van der Ryst. ISBN 0 86054 893 7

No 44   BAR S734, 1998 **Cultural Succession and Continuity in S.E. Nigeria** *Excavations in Afikpo* by V. Emenike Chikwendu. ISBN 0 86054 921 6

No 45   BAR S763, 1999 **The Emergence of Food Production in Ethiopia** by Tertia Barnett. ISBN 0 86054 971 2

No 46   BAR S768, 1999 **Sociétés préhistoriques et Mégalithes dans le Nord-Ouest de la République Centrafricaine** by Étienne Zangato. ISBN 0 86054 980 1

No 47   BAR S775, 1999 **Ethnohistoric Archaeology of the Mukogodo in North-Central Kenya** *Hunter-gatherer subsistence and the transition to pastoralism in secondary settings* by Kennedy K. Mutundu. ISBN 0 86054 990 9

No 48   BAR S782, 1999 **Échanges et contacts le long du Nil et de la Mer Rouge dans l'époque protohistorique (IIIe et IIe millénaires avant J.-C.)** *Une synthèse préliminaire* by Andrea Manzo. ISBN 1 84171 002 4. £28.00.

No 49   BAR S838, 2000 **Ethno-Archaeology in Jenné, Mali** *Craft and status among smiths, potters and masons* by Adria LaViolette. ISBN 1 84171 043 1

No 50   BAR S860, 2000 **Hunter-Gatherers and Farmers** *An enduring Frontier in the Caledon Valley, South Africa* by Carolyn R. Thorp. ISBN 1 84171 061 X. £25.00.

No 51   BAR S906, 2000 **The Kintampo Complex** *The Late Holocene on the Gambaga Escarpment, Northern Ghana* by Joanna Casey. ISBN 1 84171 202 7. £30.00.

* out of print

www.ingramcontent.com/pod-product-compliance
Lightning Source LLC
Chambersburg PA
CBHW051302270326
41926CB00030B/4697